Probability with Martingales

Probability with Martingales

David Williams
Statistical Laboratory, DPMMS
Cambridge University

CAMBRIDGE
UNIVERSITY PRESS

Published by the Press Syndicate of the University of Cambridge
The Pitt Building, Trumpington Street, Cambridge CB2 1RP
40 West 20th Street, New York, NY 10011-4211, USA
10 Stamford Road, Oakleigh, Melbourne 3166, Australia

© Cambridge University Press 1991

First published 1991
Reprinted 1992, 1994

Printed in Great Britain at the University Press, Cambridge

British Library cataloguing in publication data available

Library of Congress cataloguing in publication data

ISBN 0 521 40455 X hardback
ISBN 0 521 40605 6 paperback

Contents

PART C: CHARACTERISTIC FUNCTIONS

APPENDICES

Preface – please read!

The most important chapter in this book is *Chapter E: Exercises*. I have left the interesting things for *you* to do. You can start *now* on the 'EG' exercises, but see 'More about exercises' later in this Preface.

The book, which is essentially the set of lecture notes for a third-year undergraduate course at Cambridge, is as lively an introduction as I can manage to the rigorous theory of probability. Since much of the book is devoted to martingales, it is bound to become very lively: look at those Exercises on Chapter 10! But, of course, there is that initial plod through the measure-theoretic foundations. It must be said however that measure theory, that most arid of subjects when done for its own sake, becomes amazingly more alive when used in probability, not only because it is then applied, but also because it is immensely enriched.

You cannot avoid measure theory: an *event* in probability is a measurable set, a *random variable* is a measurable function on the sample space, the *expectation* of a random variable is its integral with respect to the probability measure; and so on. To be sure, one can take some central results from measure theory as axiomatic in the main text, giving careful proofs in appendices; and indeed that is exactly what I have done.

Measure theory for its own sake is based on the fundamental addition rule for measures. Probability theory supplements that with the multiplication rule which describes independence; and things are already looking up. But what really enriches and enlivens things is that we deal with lots of σ-algebras, not just the one σ-algebra which is the concern of measure theory.

In planning this book, I decided for every topic what things I considered just a bit too advanced, and, often with sadness, I have ruthlessly omitted them.

For a more thorough training in many of the topics covered here, see Billingsley (1979), Chow and Teicher (1978), Chung (1968), Kingman and

Taylor (1966), Laha and Rohatgi (1979), and Neveu (1965). As regards measure theory, I learnt it from Dunford and Schwartz (1958) and Halmos (1959). After reading this book, you must read the still-magnificent Breiman (1968), and, for an excellent indication of what can be done with discrete martingales, Hall and Heyde (1980).

Of course, intuition is much more important than knowledge of measure theory, and you should take every opportunity to sharpen your intuition. There is no better whetstone for this than Aldous (1989), though it is a very demanding book. For appreciating the scope of probability and for learning how to think about it, Karlin and Taylor (1981), Grimmett and Stirzaker (1982), Hall (1988), and Grimmett's recent superb book, Grimmett (1989), on percolation are strongly recommended.

More about exercises. In compiling Chapter E, which consists exactly of the homework sheet I give to the Cambridge students, I have taken into account the fact that this book, like any other mathematics book, implicitly contains a vast number of other exercises, many of which are easier than those in Chapter E. I refer of course to the exercises *you* create by reading the statement of a result, and then trying to prove it for yourself, before you read the given proof. One other point about exercises: you will, for example, surely forgive my using expectation $\mathbb{E}$ in Exercises on Chapter 4 before $\mathbb{E}$ is treated with full rigour in Chapter 6.

Acknowledgements. My first thanks must go to the students who have endured the course on which the book is based and whose quality has made me try hard to make it worthy of them; and to those, especially David Kendall, who had developed the course before it became my privilege to teach it. My thanks to David Tranah and other staff of CUP for their help in converting the course into this book. Next, I must thank Ben Garling, James Norris and Chris Rogers without whom the book would have contained more errors and obscurities; and the many readers who have kindly pointed out typing errors in earlier printings. (The many faults which surely remain in it are my responsibility.) Helen Rutherford and I typed part of the book, but the vast majority of it was typed by Sarah Shea-Simonds in a virtuoso performance worthy of Horowitz. My thanks to Helen and, most especially, to Sarah. Special thanks to my wife, Sheila, too, for all her help.

But my best thanks – and yours if you derive any benefit from the book – must go to three people whose names appear in capitals in the Index: J.L. Doob, A.N. Kolmogorov and P. Lévy: without them, there wouldn't have been much to write about, as Doob (1953) splendidly confirms.

Statistical Laboratory,   *David Williams*
Cambridge   *October 1990*

A Question of Terminology

Random variables: functions or equivalence classes?

At the level of this book, the theory would be more 'elegant' if we regarded a random variable as an *equivalence class* of measurable functions on the sample space, two functions belonging to the same equivalence class if and only if they are equal almost everywhere. Then the conditional-expectation map

$$X \mapsto \mathbb{E}(X|\mathcal{G})$$

would be a truly well-defined contraction map from $L^p(\Omega, \mathcal{F}, \mathbb{P})$ to $L^p(\Omega, \mathcal{G}, \mathbb{P})$ for $p \geq 1$; and we would not have to keep mentioning versions (representatives of equivalence classes) and would be able to avoid the endless 'almost surely' qualifications.

I have however chosen the 'inelegant' route: firstly, I prefer to work with *functions*, and confess to preferring

$$4 + 5 = 2 \bmod 7 \qquad \text{to} \qquad [4]_7 + [5]_7 = [2]_7.$$

But there is a substantive reason. I hope that this book will tempt you to progress to the much more interesting, and more important, theory where the parameter set of our process is uncountable (e.g. it may be the time-parameter set $[0, \infty)$). There, the equivalence-class formulation just will not work: the 'cleverness' of introducing quotient spaces loses the subtlety which is essential even for formulating the fundamental results on existence of continuous modifications, etc., unless one performs contortions which are hardly elegant. Even if these contortions allow one to *formulate* results, one would still have to use genuine functions to *prove* them; so where does the reality lie?!

Fran Burstall rightly insisted that I should mention that problems arise if one does not quotient out; and I have done so briefly at the end of Sections 6.7 and 6.9.

A Guide to Notation

▶ signifies something important, ▶▶ something very important, and ▶▶▶ the Martingale Convergence Theorem.

I use ':=' to signify 'is defined to equal'. This Pascal notation is particularly convenient because it can also be used in the reversed sense.

I use analysts' (as opposed to category theorists') conventions:

▶ $$\mathbb{N} := \{1, 2, 3, \ldots\} \subseteq \{0, 1, 2, \ldots\} =: \mathbb{Z}^+.$$

Everyone is agreed that $\mathbb{R}^+ := [0, \infty)$.

For a set B contained in some universal set S, I_B denotes the indicator function of B: that is $I_B : S \to \{0, 1\}$ and

$$I_B(s) := \begin{cases} 1 & \text{if } s \in B, \\ 0 & \text{otherwise.} \end{cases}$$

For $a, b \in \mathbb{R}$,

$$a \wedge b := \min(a, b), \qquad a \vee b := \max(a, b).$$

CF:characteristic function; DF: distribution function; pdf: probability density function.

σ-algebra, $\sigma(\mathcal{C})$ (1.1); $\sigma(Y_\gamma : \gamma \in \mathcal{C})$ (3.8, 3.13). π-system (1.6); d-system (A1.2).

---

a.e.:   almost everywhere (1.5)

a.s.:   almost surely (2.4)

bΣ:	the space of bounded Σ-measurable functions (3.1)
$\mathcal{B}(S)$:	the Borel σ-algebra on S, $\quad \mathcal{B} := \mathcal{B}(\mathbb{R})$ (1.2)
$C \bullet X$:	discrete stochastic integral (10.6)
$d\lambda/d\mu$:	Radon-Nikodým derivative (5.14)
$d\mathbb{Q}/d\mathbb{P}$:	Likelihood Ratio (14.13)
$\mathbb{E}(X)$:	expectation $\mathbb{E}(X) := \int_\Omega X(\omega)\mathbb{P}(d\omega)$ of X (6.3)
$\mathbb{E}(X; F)$:	$\int_F X\,d\mathbb{P}$ (6.3)
$\mathbb{E}(X\|\mathcal{G})$:	conditional expectation (9.3)
(E_n, ev):	$\liminf E_n$ (2.8)
$(E_n, \mathrm{i.o.})$:	$\limsup E_n$ (2.6)
f_X:	probability density function (pdf) of X (6.12)
$f_{X,Y}$:	joint pdf (8.3)
$f_{X\|Y}$:	conditional pdf (9.6)
F_X:	distribution function of X (3.9)
$\liminf$:	for sets, (2.8)
$\limsup$:	for sets, (2.6)
$x = \uparrow \lim x_n$:	$x_n \uparrow x$ in that $x_n \leq x_{n+1}$ ($\forall n$) and $x_n \to x$
$\log$:	natural (base e) logarithm
$\mathcal{L}_X, \Lambda_X$:	law of X (3.9)
$\mathcal{L}^p, L^p$:	Lebesgue spaces (6.7, 6.13)
Leb:	Lebesgue measure (1.8)
mΣ:	space of Σ-measurable functions (3.1)
M^T:	process M stopped at time T (10.9)
$\langle M \rangle$:	angle-brackets process (12.12)
$\mu(f)$:	integral of f with respect to μ (5.0, 5.2)
$\mu(f; A)$:	$\int_A f\,d\mu$ (5.0, 5.2)
φ_X:	CF of X (Chapter 16)
φ:	pdf of standard normal N(0,1) distribution
Φ:	DF of N(0,1) distribution
X^T:	X stopped at time T (10.9)

Chapter 0

A Branching-Process Example

(This Chapter is not essential for the remainder of the book. You can start with Chapter 1 if you wish.)

0.0. Introductory remarks

The purpose of this chapter is threefold: to take something which is probably well known to you from books such as the immortal Feller (1957) or Ross (1976), so that you start on familiar ground; to make you start to think about some of the problems involved in making the elementary treatment into rigorous mathematics; and to indicate what new results appear if one applies the somewhat more advanced theory developed in this book. We stick to one example: a branching process. This is rich enough to show that the theory has some substance.

0.1. Typical number of children, X

In our model, the number of children of a typical animal (see Notes below for some interpretations of 'child' and 'animal') is a random variable X with values in $\mathbf{Z}^+$. *We assume that*

$$\mathbf{P}(X = 0) > 0.$$

We define the *generating function f of X* as the map $f : [0,1] \rightarrow [0,1]$, where

$$f(\theta) := \mathbf{E}(\theta^X) = \sum_{k \in \mathbf{Z}^+} \theta^k \mathbf{P}(X = k).$$

Standard theorems on power series imply that, for $\theta \in [0,1]$,

$$f'(\theta) = \mathbf{E}(X\theta^{X-1}) = \sum k\theta^{k-1}\mathbf{P}(X = k)$$

and

$$\mu := \mathbf{E}(X) = f'(1) = \sum k\mathbf{P}(X = k) \le \infty.$$

Of course, $f'(1)$ is here interpreted as

$$\lim_{\theta \uparrow 1} \frac{f(\theta) - f(1)}{\theta - 1} = \lim_{\theta \uparrow 1} \frac{1 - f(\theta)}{1 - \theta},$$

since $f(1) = 1$. *We assume that*

$$\mu < \infty.$$

Notes. The first application of branching-process theory was to the question of survival of family names; and in that context, animal = man, and child = son.

In another context, 'animal' can be 'neutron', and 'child' of that neutron will signify a neutron released if and when the parent neutron crashes into a nucleus. Whether or not the associated branching process is super-critical can be a matter of real importance.

We can often find branching processes embedded in richer structures and can then use the results of this chapter to start the study of more interesting things.

For superb accounts of branching processes, see Athreya and Ney (1972), Harris (1963), Kendall (1966, 1975).

0.2. Size of nth generation, Z_n

To be a bit formal: suppose that we are given a doubly infinite sequence

(a) $$\left\{ X_r^{(m)} : m, r \in \mathbf{N} \right\}$$

of independent identically distributed random variables (IID RVs), each with the same distribution as X:

$$\mathbf{P}(X_r^{(m)} = k) = \mathbf{P}(X = k).$$

The idea is that for $n \in \mathbf{Z}^+$ and $r \in \mathbf{N}$, the variable $X_r^{(n+1)}$ represents the number of children (who will be in the $(n+1)^{\text{th}}$ generation) of the rth animal (if there is one) in the nth generation. The fundamental rule therefore is that if Z_n signifies the size of the nth generation, then

(b) $$Z_{n+1} = X_1^{(n+1)} + \cdots + X_{Z_n}^{(n+1)}.$$

We assume that $Z_0 = 1$, so that (b) gives a full recursive definition of the sequence $(Z_m : m \in \mathbf{Z}^+)$ from the sequence (a). Our first task is to calculate the distribution function of Z_n, or equivalently to find the generating function

(c) $$f_n(\theta) := \mathsf{E}(\theta^{Z_n}) = \sum \theta^k \mathsf{P}(Z_n = k).$$

0.3. Use of conditional expectations

The first main result is that for $n \in \mathbf{Z}^+$ (and $\theta \in [0,1]$)

(a) $$f_{n+1}(\theta) = f_n(f(\theta)),$$

so that for each $n \in \mathbf{Z}^+$, f_n is the n-fold composition

(b) $$f_n = f \circ f \circ \ldots \circ f.$$

Note that the 0-fold composition is by convention the identity map $f_0(\theta) = \theta$, in agreement with – indeed, forced by – the fact that $Z_0 = 1$.

To prove (a), we use – at the moment in intuitive fashion – the following very special case of the very useful *Tower Property of Conditional Expectation*:

(c) $$\mathsf{E}(U) = \mathsf{E}\mathsf{E}(U|V);$$

to find the expectation of a random variable U, first find the conditional expectation $\mathsf{E}(U|V)$ of U given V, and then find the expectation of *that*. We prove the ultimate form of (c) at a later stage.

We apply (c) with $U = \theta^{Z_{n+1}}$ and $V = Z_n$:

$$\mathsf{E}(\theta^{Z_{n+1}}) = \mathsf{E}\mathsf{E}(\theta^{Z_{n+1}}|Z_n).$$

Now, for $k \in \mathbf{Z}^+$, the conditional expectation of $\theta^{Z_{n+1}}$ given that $Z_n = k$ satisfies

(d) $$\mathsf{E}(\theta^{Z_{n+1}}|Z_n = k) = \mathsf{E}(\theta^{X_1^{(n+1)}+\cdots+X_k^{(n+1)}}|Z_n = k).$$

But Z_n is constructed from variables $X_s^{(r)}$ with $r \leq n$, and so Z_n is independent of $X_1^{(n+1)}, \ldots, X_k^{(n+1)}$. The conditional expectation given $Z_n = k$ in the right-hand term in (d) must therefore agree with the absolute expectation

(e) $$\mathsf{E}(\theta^{X_1^{(n+1)}} \ldots \theta^{X_k^{(n+1)}}).$$

But the expression at (e) is an *expectation of the product of independent random variables* and as part of the family of '*Independence means multiply*' results, we know that this expectation of a product may be rewritten as the product of expectations. Since (for every n and r)

$$\mathbb{E}(\theta^{X_r^{(n+1)}}) = f(\theta),$$

we have proved that

$$\mathbb{E}(\theta^{Z_{n+1}} | Z_n = k) = f(\theta)^k,$$

and this is what it means to say that

$$\mathbb{E}(\theta^{Z_{n+1}} | Z_n) = f(\theta)^{Z_n}.$$

[If V takes only integer values, then when $V = k$, the conditional expectation $\mathbb{E}(U|V)$ of U given V is equal to the conditional expectation $\mathbb{E}(U|V = k)$ of U given that $V = k$. (Sounds reasonable!)] Property (c) now yields

$$\mathbb{E}\theta^{Z_{n+1}} = \mathbb{E}f(\theta)^{Z_n},$$

and, since

$$\mathbb{E}(\alpha^{Z_n}) = f_n(\alpha), \qquad\qquad \square$$

result (a) is proved.

Independence and *conditional expectations* are two of the main topics in this course.

0.4. Extinction probability, π

Let $\pi_n := \mathbb{P}(Z_n = 0)$. Then $\pi_n = f_n(0)$, so that, by (0.3,b),

(a) $$\pi_{n+1} = f(\pi_n).$$

Measure theory confirms our intuition about the extinction probability:

(b) $$\pi := \mathbb{P}(Z_m = 0 \text{ for some } m) = \uparrow \lim \pi_n.$$

Because f is continuous, it follows from (a) that

(c) $$\pi = f(\pi).$$

The function f is analytic on (0,1), and is non-decreasing and convex (of non-decreasing slope). Also, $f(1) = 1$ and $f(0) = \mathbb{P}(X = 0) > 0$. The slope $f'(1)$ of f at 1 is $\mu = \mathbb{E}(X)$. The celebrated pictures opposite now make the following Theorem obvious.

THEOREM

If $\mathbb{E}(X) > 1$, then the extinction probability π is the unique root of the equation $\pi = f(\pi)$ which lies strictly between 0 and 1. If $\mathbb{E}(X) \leq 1$, then $\pi = 1$.

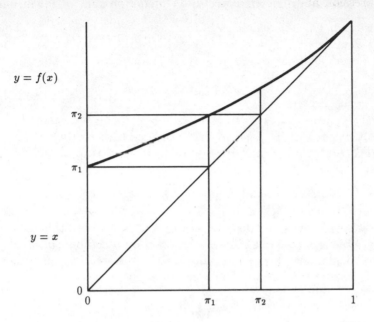

$y = f(x)$

π_2

π_1

$y = x$

0

0   π_1 π_2 1

Case 1: subcritical, $\mu = f'(1) < 1$. Clearly, $\pi = 1$.
The *critical case* $\mu = 1$ has a similar picture.

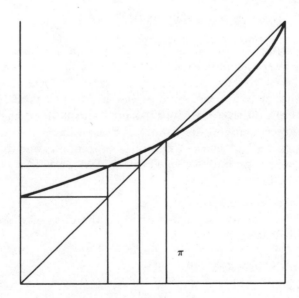

π

Case 2: supercritical, $\mu = f'(1) > 1$. Now, $\pi < 1$.

0.5. Pause for thought: measure

Now that we have finished revising what introductory courses on probability theory say about branching-process theory, let us think about why we must find a more precise language. To be sure, the claim at (0.4,b) that

(a) $$\pi =\uparrow \lim \pi_n$$

is intuitively plausible, but how could one *prove* it? We certainly cannot prove it at present because we have no means of stating with pure-mathematical precision what it is supposed to mean. Let us discuss this further.

Back in Section 0.2, we said 'Suppose that we are given a doubly infinite sequence $\{X_r^{(m)} : m, r \in \mathsf{N}\}$ of independent identically distributed random variables each with the same distribution as X'. What does this mean? A random variable is a (certain kind of) function on a sample space Ω. We could follow elementary theory in taking Ω to be the set of all outcomes, in other words, taking Ω to be the Cartesian product

$$\Omega = \prod_{r,s} \mathsf{Z}^+,$$

the typical element ω of Ω being

$$\omega = (\omega_s^{(r)} : r \in \mathsf{N}, s \in \mathsf{N}),$$

and then setting $X_s^{(r)}(\omega) = \omega_s^{(r)}$. Now Ω is an uncountable set, so that we are outside the 'combinatorial' context which makes sense of π_n in the elementary theory. Moreover, if one assumes the Axiom of Choice, one can *prove* that it is impossible to assign to *all* subsets of Ω a probability satisfying the 'intuitively obvious' axioms and making the X's IID RVs with the correct common distribution. So, we have to know that the set of ω corresponding to the event 'extinction occurs' *is* one to which one can uniquely assign a probability (which will then provide a definition of π). Even then, we have to prove (a).

Example. Consider for a moment what is in some ways a bad attempt to construct a 'probability theory'. Let $\mathcal{C}$ be the class of subsets C of N for which the 'density'

$$\rho(C) := \lim_{n\uparrow\infty} n^{-1}\#\{k : 1 \leq k \leq n; k \in C\}$$

exists. Let $C_n := \{1, 2, \ldots, n\}$. Then $C_n \in \mathcal{C}$ and $C_n \uparrow \mathsf{N}$ in the sense that $C_n \subseteq C_{n+1}, \forall n$ and also $\bigcup C_n = \mathsf{N}$. However, $\rho(C_n) = 0, \forall n$, but $\rho(\mathsf{N}) = 1$.

Hence the logic which will allow us correctly to deduce (a) from the fact that

$$\{Z_n = 0\} \uparrow \{\text{extinction occurs}\}$$

fails for the $(\mathbf{N}, \mathcal{C}, \rho)$ set-up: $(\mathbf{N}, \mathcal{C}, \rho)$ is not 'a probability triple'. $\qquad \square$

There *are* problems. Measure theory resolves them, but provides a huge bonus in the form of much deeper results such as the Martingale Convergence Theorem which we now take a first look at – at an intuitive level, I hasten to add.

0.6. Our first martingale

Recall from (0.2,b) that

$$Z_{n+1} = X_1^{(n+1)} + \cdots + X_{Z_n}^{(n+1)},$$

where the $X^{(n+1)}$ variables are independent of the values $Z_1, Z_2, \ldots, Z_n$. It is clear from this that

$$\mathbf{P}(Z_{n+1} = j | Z_0 = i_0, Z_1 = i_1, \ldots, Z_n = i_n) = \mathbf{P}(Z_{n+1} = j | Z_n = i_n),$$

a result which you will probably recognize as stating that the process $Z = (Z_n : n \geq 0)$ *is a Markov chain.* We therefore have

$$\mathbf{E}(Z_{n+1} | Z_0 = i_0, Z_1 = i_1, \ldots, Z_n = i_n) = \sum_j j \mathbf{P}(Z_{n+1} = j | Z_n = i_n)$$

$$= \mathbf{E}(Z_{n+1} | Z_n = i_n),$$

or, in a condensed and better notation,

(a) $$\mathbf{E}(Z_{n+1} | Z_0, Z_1, \ldots, Z_n) = \mathbf{E}(Z_{n+1} | Z_n).$$

Of course, it is intuitively obvious that

(b) $$\mathbf{E}(Z_{n+1} | Z_n) = \mu Z_n,$$

because each of the Z_n animals in the nth generation has on average μ children. We can confirm result (b) by differentiating the result

$$\mathbf{E}(\theta^{Z_{n+1}} | Z_n) = f(\theta)^{Z_n}$$

with respect to θ and setting $\theta = 1$.

Now define

(c) $$M_n := Z_n/\mu^n, \qquad n \geq 0.$$

Then

$$E(M_{n+1}|Z_0, Z_1, \ldots, Z_n) = M_n,$$

which exactly says that

(d) *M is a martingale relative to the Z process.*

Given the history of Z up to stage n, the next value M_{n+1} of M is on average what it is now: M is 'constant on average' in this very sophisticated sense of conditional expectation given 'past' and 'present'. The true statement

(e) $$E(M_n) = 1, \quad \forall n$$

is of course infinitely cruder.

A statement S is said to be true **almost surely** (a.s.) or **with probability 1** if (surprise, surprise!)

$$P(S \text{ is true}) = 1.$$

Because our martingale M is *non-negative* ($M_n \geq 0, \forall n$), the **Martingale Convergence Theorem** implies that *it is almost surely true that*

(f) $$M_\infty := \lim M_n \quad exists.$$

Note that if $M_\infty > 0$ for some outcome (which can happen with positive probability only when $\mu > 1$), then the statement

$$Z_n/\mu^n \to M_\infty \qquad \text{(a.s.)}$$

is a precise formulation of 'exponential growth'. A particularly fascinating question is: *suppose that $\mu > 1$; what is the behaviour of Z conditional on the value of M_∞?*

0.7. Convergence (or not) of expectations

We know that $M_\infty := \lim M_n$ exists with probability 1, and that $E(M_n) = 1$, $\forall n$. We might be tempted to believe that $E(M_\infty) = 1$. However, we already know that if $\mu \leq 1$, then, almost surely, the process dies out and M_n is eventually 0. Hence

(a) $$\text{if } \mu \leq 1, \text{ then } M_\infty = 0 \text{ (a.s.) } and$$

$$0 = E(M_\infty) \neq \lim E(M_n) = 1.$$

This is an excellent example to keep in mind when we come to study *Fatou's Lemma*, valid for any sequence (Y_n) of non-negative random variables:

$$\mathsf{E}(\liminf Y_n) \leq \liminf \mathsf{E}(Y_n).$$

What is 'going wrong' at (a) is that (when $\mu \leq 1$) for large n, the chances are that M_n will be large if M_n is not 0 and, very roughly speaking, this large value times its small probability will keep $\mathsf{E}(M_n)$ at 1. See the concrete examples in Section 0.9.

Of course, it is very important to know when

(b) $$\lim \mathsf{E}(\cdot) = \mathsf{E}(\lim \cdot),$$

and we do spend quite a considerable time studying this. The best general theorems are rarely good enough to get the best results for concrete problems, as is evidenced by the fact that

(c) $\mathsf{E}(M_\infty) = 1$ *if and only if* both $\mu > 1$ *and* $\mathsf{E}(X \log X) < \infty$,

where X is the typical number of children. Of course $0 \log 0 = 0$. If $\mu > 1$ and $\mathsf{E}(X \log X) = \infty$, then, even though the process may not die out, $M_\infty = 0$, a.s.

0.8. Finding the distribution of M_∞

Since $M_n \to M_\infty$ (a.s.), it is obvious that for $\lambda > 0$,

$$\exp(-\lambda M_n) \to \exp(-\lambda M_\infty) \qquad \text{(a.s.)}$$

Now since each $M_n \geq 0$, the whole sequence $(\exp(-\lambda M_n))$ is bounded in absolute value by the constant 1, independently of the outcome of our experiment. The *Bounded Convergence Theorem* says that we *can* now assert what we would wish:

(a) $$\mathsf{E}\exp(-\lambda M_\infty) = \lim \mathsf{E}\exp(-\lambda M_n).$$

Since $M_n = Z_n/\mu^n$ and $\mathsf{E}(\theta^{Z_n}) = f_n(\theta)$, we have

(b) $$\mathsf{E}\exp(-\lambda M_n) = f_n(\exp(-\lambda/\mu^n)),$$

so that, in principle (if very rarely in practice), we can calculate the left-hand side of (a). However, for a non-negative random variable Y, *the distribution function* $y \mapsto \mathsf{P}(Y \leq y)$ *is completely determined by the map*

$$\lambda \mapsto \mathsf{E}\exp(-\lambda Y) \quad on \quad (0, \infty).$$

Hence, in principle, we can find the distribution of M_∞.

We have seen that the real problem is to calculate the function

$$L(\lambda) := \mathsf{E} \exp(-\lambda M_\infty).$$

Using (b), the fact that $f_{n+1} = f \circ f_n$, and the continuity of L (another consequence of the Bounded Convergence Theorem), you can immediately establish the functional equation:

(c) $$L(\lambda\mu) = f(L(\lambda)).$$

0.9. Concrete example

This concrete example is just about the only one in which one can calculate everything explicitly, but, in the way of mathematics, it is useful in many contexts.

We take the 'typical number of children' X to have a *geometric distribution*:

(a) $$\mathsf{P}(X = k) = pq^k \quad (k \in \mathbf{Z}^+),$$

where

$$0 < p < 1, \quad q := 1 - p.$$

Then, as you can easily check,

(b) $$f(\theta) = \frac{p}{1 - q\theta}, \quad \mu = \frac{q}{p},$$

and

$$\pi = \begin{cases} p/q & \text{if } q > p, \\ 1 & \text{if } q \le p. \end{cases}$$

To calculate $f \circ f \circ \ldots \circ f$, we use a device familiar from the geometry of the upper half-plane. If

$$G = \begin{pmatrix} g_{11} & g_{12} \\ g_{21} & g_{22} \end{pmatrix}$$

is a non-singular 2×2 matrix, define the fractional linear transformation:

(c) $$G(\theta) = \frac{g_{11}\theta + g_{12}}{g_{21}\theta + g_{22}}.$$

Then you can check that if H is another such matrix, then

$$G(H(\theta)) = (GH)(\theta),$$

so that composition of fractional linear transformations corresponds to matrix multiplication.

Suppose that $p \neq q$. Then, by the $S^{-1}AS = \Lambda$ method, for example, we find that the nth power of the matrix corresponding to f is

$$\begin{pmatrix} 0 & p \\ -q & 1 \end{pmatrix}^n = (q-p)^{-1} \begin{pmatrix} 1 & p \\ 1 & q \end{pmatrix} \begin{pmatrix} p^n & 0 \\ 0 & q^n \end{pmatrix} \begin{pmatrix} q & -p \\ -1 & 1 \end{pmatrix},$$

so that

(d)
$$f_n(\theta) = \frac{p\mu^n(1-\theta) + q\theta - p}{q\mu^n(1-\theta) + q\theta - p}.$$

If $\mu = q/p \leq 1$, then $\lim_n f_n(\theta) = 1$, corresponding to the fact that the process dies out.

Suppose now that $\mu > 1$. Then you can easily check that, for $\lambda > 0$,

$$L(\lambda) := \mathbf{E}\exp(-\lambda M_\infty) = \lim f_n(\exp(-\lambda/\mu^n))$$
$$= \frac{p\lambda + q - p}{q\lambda + q - p}$$
$$= \pi e^{-\lambda \cdot 0} + \int_0^\infty (1-\pi)^2 e^{-\lambda x} e^{-(1-\pi)x} dx,$$

from which we deduce that

$$\mathbf{P}(M_\infty = 0) = \pi,$$

and

$$\mathbf{P}(x < M_\infty < x + dx) = (1-\pi)^2 e^{-(1-\pi)x} dx \quad (x > 0),$$

or, better,

$$\mathbf{P}(M_\infty > x) = (1-\pi)e^{-(1-\pi)x} \quad (x > 0).$$

Suppose that $\mu < 1$. In this case, it is interesting to ask: what is the distribution of Z_n conditioned by $Z_n \neq 0$? We find that

$$\mathbf{E}(\theta^{Z_n}|Z_n \neq 0) = \frac{f_n(\theta) - f_n(0)}{1 - f_n(0)} = \frac{\alpha_n \theta}{1 - \beta_n \theta},$$

where

$$\alpha_n = \frac{p - q}{p - q\mu^n}, \qquad \beta_n = \frac{q - q\mu^n}{p - q\mu^n},$$

so $0 < \alpha_n < 1$ and $\alpha_n + \beta_n = 1$. As $n \to \infty$, we see that

$$\alpha_n \to 1 - \mu, \quad \beta_n \to \mu,$$

so (this is justified)

(e) $$\lim_{n \to \infty} \mathbf{P}(Z_n = k | Z_n \neq 0) = (1 - \mu)\mu^{k-1} \qquad (k \in \mathbf{N}).$$

Suppose that $\mu = 1$. You can show by induction that

$$f_n(\theta) = \frac{n - (n-1)\theta}{(n+1) - n\theta},$$

and that

$$\mathbf{E}(e^{-\lambda Z_n / n} | Z_n \neq 0) \to 1/(1 + \lambda),$$

corresponding to

(f) $$\mathbf{P}(Z_n / n > x | Z_n \neq 0) \to e^{-x}, \qquad x > 0.$$

'The Fatou factor'

We know that when $\mu \leq 1$, we have $\mathbf{E}(M_n) = 1$, $\forall n$, but $\mathbf{E}(M_\infty) = 0$. Can we get some insight into this?

First consider the case when $\mu < 1$. Result (e) makes it plausible that for large n,

$$\mathbf{E}(Z_n | Z_n \neq 0) \text{ is roughly } (1 - \mu) \sum k\mu^{k-1} = 1/(1 - \mu).$$

We know that

$$\mathbf{P}(Z_n \neq 0) = 1 - f_n(0) \text{ is roughly } (1 - \mu)\mu^n,$$

so we should have (roughly)

$$\mathbf{E}(M_n) = \mathbf{E}\left(\frac{Z_n}{\mu^n} \middle| Z_n \neq 0\right) \mathbf{P}(Z_n \neq 0)$$

$$= \frac{1}{(1 - \mu)\mu^n}(1 - \mu)\mu^n = 1,$$

which might help explain how the 'balance' $\mathbf{E}(M_n) = 1$ is achieved by big values times small probabilities.

Now consider the case when $\mu = 1$. Then

$$\mathbf{P}(Z_n \neq 0) = 1/(n+1),$$

and, from (f), Z_n/n conditioned by $Z_n \neq 0$ is roughly exponential with mean 1, so that $M_n = Z_n$ conditioned by $Z_n \neq 0$ is on average of size about n, the correct order of magnitude for balance.

Warning. We have just been using for 'correct intuitive explanations' exactly the type of argument which might have misled us into thinking that $\mathbf{E}(M_\infty) = 1$ in the first place. But, of course, the result

$$\mathbf{E}(M_n) = \mathbf{E}(M_n|Z_n \neq 0)\mathbf{P}(Z_n \neq 0) = 1$$

is a matter of obvious *fact*.

PART A: FOUNDATIONS

Chapter 1
Measure Spaces

1.0. Introductory remarks

Topology is about *open* sets. The characterizing property of a *continuous* function f is that the inverse image $f^{-1}(G)$ of an open set G is open.

Measure theory is about *measurable* sets. The characterizing property of a *measurable* function f is that the inverse image $f^{-1}(A)$ of any measurable set is measurable.

In topology, one axiomatizes the notion of 'open set', insisting in particular that the union of *any* collection of open sets is open, and that the intersection of a *finite* collection of open sets is open.

In measure theory, one axiomatizes the notion of 'measurable set', insisting that the union of a *countable* collection of measurable sets is measurable, and that the intersection of a *countable* collection of measurable sets is also measurable. Also, the complement of a measurable set must be measurable, and the whole space must be measurable. Thus the measurable sets form a σ-algebra, a structure stable (or 'closed') under countably many set operations. Without the insistence that 'only countably many operations are allowed', measure theory would be self-contradictory – a point lost on certain philosophers of probability.

The probability that a point chosen at random on the surface of the unit sphere S^2 in $\mathbf{R}^3$ falls into the subset F of S^2 is just the area of F divided by the total area 4π. What could be easier?

However, Banach and Tarski showed (see Wagon (1985)) that if the Axiom of Choice is assumed, as it is throughout conventional mathematics, then there exists a subset F of the unit sphere S^2 in $\mathbf{R}^3$ such that for

$3 \leq k < \infty$ (and even for $k = \infty$), S^2 is the disjoint union of k exact copies of F:

$$S^2 = \bigcup_{i=1}^{k} \tau_i^{(k)} F,$$

where each $\tau_i^{(k)}$ is a rotation. If F has an 'area', then that area must simultaneously be $4\pi/3, 4\pi/4, \ldots, 0$. The only conclusion is that the set F is *non-measurable* (not *Lebesgue* measurable): it is so complicated that one cannot assign an area to it. Banach and Tarski have not broken the Law of Conservation of Area: they have simply operated outside its jurisdiction.

Remarks. (i) Because every rotation τ has a fixed point x on S^2 such that $\tau(x) = x$, it is not possible to find a subset A of S^2 and a rotation τ such that $A \cup \tau(A) = S^2$ and $A \cap \tau(A) = \emptyset$. So, we could not have taken $k = 2$.

(ii) Banach and Tarski even proved that given any two bounded subsets A and B of $\mathbf{R}^3$ each with non-empty interior, it is possible to decompose A into a certain finite number n of disjoint pieces $A = \bigcup_{i=1}^{n} A_i$ and B into the same number n of disjoint pieces $B = \bigcup_{i=1}^{n} B_i$, in such a way that, for each i, A_i is Euclid-congruent to B_i!!! So, we can disassemble A and rebuild it as B.

(iii) Section A1.1 (optional!) in the appendix to this chapter gives an Axiom-of-Choice construction of a non-measurable subset of S^1.

This chapter introduces

$$\sigma\text{-}algebras,\ \pi\text{-}systems,\ \text{and}\ measures$$

and emphasizes *monotone-convergence properties of measures*. We shall see in later chapters that, although not all sets are measurable, it is always the case for probability theory that enough sets are measurable.

1.1. Definitions of algebra, σ-algebra

Let S be a set.

Algebra on S

A collection Σ_0 of subsets of S is called an *algebra on S* (or algebra of subsets of S) if
- (i) $S \in \Sigma_0$,
- (ii) $F \in \Sigma_0 \quad \Rightarrow \quad F^c := S \backslash F \in \Sigma_0$,
- (iii) $F, G \in \Sigma_0 \quad \Rightarrow \quad F \cup G \in \Sigma_0$.

[Note that $\emptyset = S^c \in \Sigma_0$ and

$$F, G \in \Sigma_0 \quad \Rightarrow \quad F \cap G = (F^c \cup G^c)^c \in \Sigma_0.]$$

Thus, an algebra on S is a family of subsets of S stable under finitely many set operations.

Exercise (optional). Let $\mathcal{C}$ be the class of subsets C of $\mathbb{N}$ for which the 'density'

$$\lim_{m\uparrow\infty} m^{-1}\#\{k : 1 \leq k \leq m; k \in C\}$$

exists. We might like to think of this density (if it exists) as 'the probability that a number chosen at random belongs to C'. But there are many reasons why this does not conform to a proper probability theory. (We saw one in Section 0.5.) For example, you should find elements F and G in $\mathcal{C}$ for which $F \cap G \notin \mathcal{C}$.

Note on terminology ('algebra versus field'). An algebra in our sense is a true algebra in the algebraists' sense with $\cap$ as product, and symmetric difference

$$A \triangle B := (A \cup B)\backslash(A \cap B)$$

as 'sum', the underlying field of the algebra being the field with 2 elements. (This is why we prefer 'algebra of subsets' to 'field of subsets': there is no way that an algebra of subsets is a field in the algebraists' sense - unless Σ_0 is *trivial*, that is, $\Sigma_0 = \{S, \emptyset\}$.)

σ-algebra on S

A collection Σ of subsets of S is called a σ-*algebra on S* (or σ-algebra of subsets of S) if Σ is an algebra on S such that *whenever* $F_n \in \Sigma$ $(n \in \mathbb{N})$, *then*

$$\bigcup_n F_n \in \Sigma.$$

[Note that if Σ is a σ-algebra on S and $F_n \in \Sigma$ for $n \in \mathbb{N}$, then

$$\bigcap_n F_n = \left(\bigcup_n F_n^c\right)^c \in \Sigma.]$$

Thus, a σ-algebra on S is a family of subsets of S 'stable under any countable collection of set operations'.

Note. Whereas it is usually possible to write in 'closed form' the typical element of many of the algebras of sets which we shall meet (see Section 1.8 below for a first example), it is usually impossible to write down the typical element of a σ-algebra. This is the reason for our concentrating where possible on the much simpler 'π-systems'.

Measurable space

A pair (S, Σ), where S is a set and Σ is a σ-algebra on S, is called a *measurable space*. An element of Σ is called a Σ-measurable subset of S.

$\sigma(\mathcal{C})$, σ-algebra generated by a class $\mathcal{C}$ of subsets

Let $\mathcal{C}$ be a class of subsets of S. Then $\sigma(\mathcal{C})$, the *σ-algebra generated by $\mathcal{C}$*, is the smallest σ-algebra Σ on S such that $\mathcal{C} \subseteq \Sigma$. It is the intersection of all σ-algebras on S which have $\mathcal{C}$ as a subclass. (Obviously, the class of *all* subsets of S is a σ-algebra which extends $\mathcal{C}$.)

1.2. Examples. Borel σ-algebras, $\mathcal{B}(S)$, $\mathcal{B} = \mathcal{B}(\mathbf{R})$

Let S be a topological space.

$\mathcal{B}(S)$

$\mathcal{B}(S)$, the Borel σ-algebra on S, is the σ-algebra generated by the family of open subsets of S. With slight abuse of notation,

$$\mathcal{B}(S) := \sigma(\text{open sets}).$$

$\mathcal{B} := \mathcal{B}(\mathbf{R})$

It is standard shorthand that $\mathcal{B} := \mathcal{B}(\mathbf{R})$.

The σ-algebra $\mathcal{B}$ is the most important of all σ-algebras. Every subset of $\mathbf{R}$ which you meet in everyday use is an element of $\mathcal{B}$; and indeed it is difficult (but possible!) to find a subset of $\mathbf{R}$ constructed explicitly (without the Axiom of Choice) which is not in $\mathcal{B}$.

Elements of $\mathcal{B}$ can be quite complicated. However, the collection

$$\pi(\mathbf{R}) := \{(-\infty, x] : x \in \mathbf{R}\}$$

(not a standard notation) is very easy to understand, and it is often the case that all we need to know about $\mathcal{B}$ is that

(a) $$\mathcal{B} = \sigma(\pi(\mathbf{R})).$$

Proof of (a). For each x in $\mathbf{R}$, $(-\infty, x] = \bigcap_{n \in \mathbf{N}}(-\infty, x + n^{-1})$, so that as a *countable* intersection of open sets, the set $(-\infty, x]$ is in $\mathcal{B}$.

All that remains to be proved is that every open subset G of $\mathbf{R}$ is in $\sigma(\pi(\mathbf{R}))$. But every such G is a *countable* union of open intervals, so we need only show that, for $a, b \in \mathbf{R}$ with $a < b$,

$$(a, b) \in \sigma(\pi(\mathbf{R})).$$

But, for any u with $u > a$,

$$(a, u] = (-\infty, u] \cap (-\infty, a]^c \in \sigma(\pi(\mathbf{R})),$$

and since, for $\varepsilon = \frac{1}{2}(b - a)$,

$$(a, b) = \bigcup_n (a, b - \varepsilon n^{-1}],$$

we see that $(a, b) \in \sigma(\pi(\mathbf{R}))$, and the proof is complete. $\qquad\square$

1.3. Definitions concerning set functions

Let S be a set, let Σ_0 be an algebra on S, and let μ_0 be a non-negative *set function*

$$\mu_0 : \Sigma_0 \to [0, \infty].$$

Additive

Then μ_0 is called *additive* if $\mu_0(\emptyset) = 0$ and, for $F, G \in \Sigma_0$,

$$F \cap G = \emptyset \quad \Rightarrow \quad \mu_0(F \cup G) = \mu_0(F) + \mu_0(G).$$

Countably additive

The map μ_0 is called *countably additive* (or σ-additive) if $\mu(\emptyset) = 0$ and whenever $(F_n : n \in \mathbb{N})$ is a sequence of disjoint sets in Σ_0 with union $F = \bigcup F_n$ in Σ_0 (note that this is an assumption since Σ_0 need not be a σ-algebra), then

$$\mu_0(F) = \sum_n \mu_0(F_n).$$

Of course (why?), a countably additive set function is additive.

1.4. Definition of measure space

Let (S, Σ) be a measurable space, so that Σ is a σ-algebra on S.
  A map

$$\mu : \Sigma \to [0, \infty].$$

is called a *measure* on (S, Σ) if μ is countably additive. The triple (S, Σ, μ) is then called a *measure space*.

1.5. Definitions concerning measures

Let (S, Σ, μ) be a measure space. Then μ (or indeed the measure space (S, Σ, μ)) is called

finite
  if $\mu(S) < \infty$,

σ-finite
  if there is a sequence $(S_n : n \in \mathbb{N})$ of elements of Σ such that

$$\mu(S_n) < \infty \ (\forall n \in \mathbb{N}) \text{ and } \bigcup S_n = S.$$

Warning. Intuition is usually OK for finite measures, and adapts well for σ-finite measures. However, measures which are not σ-finite can be crazy; fortunately, there are no such measures in this book.

Probability measure, probability triple

Our measure μ is called a *probability measure* if

$$\mu(S) = 1,$$

and (S, Σ, μ) is then called a *probability triple*.

μ-null element of Σ, almost everywhere (a.e.)

An element F of Σ is called *μ-null* if $\mu(F) = 0$. A statement $\mathcal{S}$ about points s of S is said to hold *almost everywhere* (a.e.) if

$$F := \{s : \mathcal{S}(s) \text{ is false}\} \in \Sigma \text{ and } \mu(F) = 0.$$

1.6. LEMMA. Uniqueness of extension, π-systems

Moral: σ-algebras are 'difficult', but π-systems are 'easy'; so we aim to work with the latter.

▶(a) *Let S be a set. Let $\mathcal{I}$ be a π-system on S, that is, a family of subsets of S* **stable under finite intersection***:*

$$I_1, I_2 \in \mathcal{I} \quad \Rightarrow \quad I_1 \cap I_2 \in \mathcal{I}.$$

Let $\Sigma := \sigma(\mathcal{I})$. Suppose that μ_1 and μ_2 are measures on (S, Σ) such that $\mu_1(S) = \mu_2(S) < \infty$ and $\mu_1 = \mu_2$ on $\mathcal{I}$. Then

$$\mu_1 = \mu_2 \text{ on } \Sigma.$$

▶▶(b) **Corollary. If two probability measures agree on a π-system, then they agree on the σ-algebra generated by that π-system.**

The example $\mathcal{B} = \sigma(\pi(\mathbf{R}))$ is of course the most important example of the $\Sigma = \sigma(\mathcal{I})$ in the theorem.

This result will play an important rôle. Indeed, it will be applied more frequently than will the celebrated existence result in Section 1.7. Because of this, the proof of Lemma 1.6 given in Sections A1.2-1.4 of the appendix to this chapter should perhaps be consulted – but read the remainder of this chapter first.

1.7. THEOREM. Carathéodory's Extension Theorem

▶   *Let S be a set, let Σ_0 be an algebra on S, and let*

$$\Sigma := \sigma(\Sigma_0).$$

If μ_0 is a countably additive map $\mu_0 : \Sigma_0 \to [0, \infty]$, then there exists a measure μ on (S, Σ) such that

$$\mu = \mu_0 \text{ on } \Sigma_0.$$

If $\mu_0(S) < \infty$, then, by Lemma 1.6, this extension is unique – an algebra is a π-system!

In a sense, this result should have more ▶ signs than any other, for without it we could not construct any interesting models. However, once we have our model, we make no further use of the theorem.

The proof of this result given in Sections A1.5-1.8 of the appendix is there for completeness. It will do no harm to assume the result for this course. Let us now see how the theorem is *used*.

1.8. Lebesgue measure Leb on $((0,1], \mathcal{B}(0,1])$

Let $S = (0,1]$. For $F \subseteq S$, say that $F \in \Sigma_0$ if F may be written as a finite union

$$(*) \qquad\qquad F = (a_1, b_1] \cup \ldots \cup (a_r, b_r]$$

where $r \in \mathbb{N}$, $0 \le a_1 \le b_1 \le \cdots \le a_r \le b_r \le 1$. Then Σ_0 is an algebra on $(0,1]$ and

$$\Sigma := \sigma(\Sigma_0) = \mathcal{B}(0,1].$$

(We write $\mathcal{B}(0,1]$ instead of $\mathcal{B}((0,1])$.) For F as at $(*)$, let

$$\mu_0(F) = \sum_{k \le r} (b_k - a_k).$$

Then μ_0 is well-defined and additive on Σ_0 (this is easy). Moreover, μ_0 is *countably* additive on Σ_0. (This is not trivial. See Section A1.9.) Hence, by Theorem 1.7, there exists a unique measure μ on $((0,1], \mathcal{B}(0,1])$ extending μ_0 on Σ_0. This measure μ is called *Lebesgue measure* on $((0,1], \mathcal{B}(0,1])$ or (loosely) Lebesgue measure on (0,1]. We shall often denote μ by Leb. Lebesgue measure (still denoted by Leb) on $([0,1], \mathcal{B}[0,1])$ is of course obtained by a trivial modification, the set $\{0\}$ having Lebesgue measure 0. Of course, Leb makes precise the concept of length.

In a similar way, we can construct (σ-finite) Lebesgue measure (which we also denote by Leb) on $\mathbb{R}$ (more strictly, on $(\mathbb{R}, \mathcal{B}(\mathbb{R}))$).

1.9. LEMMA. Elementary inequalities

Let (S, Σ, μ) be a measure space. Then

(a) $\mu(A \cup B) \leq \mu(A) + \mu(B)$ $(A, B \in \Sigma)$,

▶(b) $\mu(\bigcup_{i \leq n} F_i) \leq \sum_{i \leq n} \mu(F_i)$ $(F_1, F_2, \ldots, F_n \in \Sigma)$.

Furthermore, if $\mu(S) < \infty$, then

(c) $\mu(A \cup B) = \mu(A) + \mu(B) - \mu(A \cap B)$ $(A, B \in \Sigma)$,

(d) **(inclusion-exclusion formula):** *for $F_1, F_2, \ldots, F_n \in \Sigma$,*

$$\mu(\bigcup_{i \leq n} F_i) = \sum_{i \leq n} \mu(F_i) - \sum\sum_{i < j \leq n} \mu(F_i \cap F_j)$$

$$+ \sum\sum\sum_{i<j<k\leq n} \mu(F_i \cap F_j \cap F_k) - \cdots + (-1)^{n-1} \mu(F_1 \cap F_2 \cap \ldots \cap F_n),$$

successive partial sums alternating between over- and under-estimates.

You will surely have seen some version of these results previously. Result (c) is obvious because $A \cup B$ is the disjoint union $A \cup (B \backslash (A \cap B))$. But (c)⇒(a)⇒(b) – check that 'infinities do not matter'. You can deduce (d) from (c) by induction, but, as we shall see later, the neat way to prove (d) is by integration.

1.10. LEMMA. Monotone-convergence properties of measures

These results are often needed for making things rigorous. (Peep ahead to the 'Monkey typing Shakespeare' Section 4.9.) Again, let (S, Σ, μ) be a measure space.

▶(a) **If** $F_n \in \Sigma$ $(n \in \mathbb{N})$ **and** $F_n \uparrow F$, **then** $\mu(F_n) \uparrow \mu(F)$.

Notes. $F_n \uparrow F$ means: $F_n \subseteq F_{n+1}$ $(\forall n \in \mathbb{N})$, $\bigcup F_n = F$. Result (a) is *the* fundamental property of measure.

Proof of (a). Write $G_1 := F_1$, $G_n := F_n \backslash F_{n-1}$ $(n \geq 2)$. Then the sets G_n $(n \in \mathbb{N})$ are *disjoint*, and

$$\mu(F_n) = \mu(G_1 \cup G_2 \cup \ldots \cup G_n) = \sum_{k \leq n} \mu(G_k) \uparrow \sum_{k < \infty} \mu(G_k) = \mu(F). \quad \square$$

Application. In a proper formulation of the branching-process example of Chapter 0,

$$\{Z_n = 0\} \uparrow \{\text{extinction occurs}\}, \text{ so that } \pi_n \uparrow \pi.$$

(A proper formulation of the branching-process example *will* be given later.)

▶(b) **If $G_n \in \Sigma$, $G_n \downarrow G$ and $\mu(G_k) < \infty$ for some k, then $\mu(G_n) \downarrow \mu(G)$.**

Proof of (b). For $n \in \mathbb{N}$, let $F_n := G_k \backslash G_{k+n}$, and now apply part (a). $\qquad \square$

Example - *to indicate what can 'go wrong'.* For $n \in \mathbb{N}$, let

$$H_n := (n, \infty).$$

Then $\mathrm{Leb}(H_n) = \infty, \forall n$, but $H_n \downarrow \emptyset$.

▶(c) *The union of a countable number of μ-null sets is μ-null.*

This is a trivial corollary of results (1.9,b) and (1.10,a).

1.11. Example/Warning

Let (S, Σ, μ) be $([0,1], \mathcal{B}[0,1], \mathrm{Leb})$. Let $\varepsilon(k)$ be a sequence of strictly positive numbers such that $\varepsilon(k) \downarrow 0$. For a single point x of S, we have

(a) $$\{x\} \subseteq (x - \varepsilon(k), x + \varepsilon(k)) \cap S,$$

so that for every k, $\mu(\{x\}) < 2\varepsilon(k)$, and so $\mu(\{x\}) = 0$. That $\{x\}$ is $\mathcal{B}(S)$-measurable follows because $\{x\}$ is the intersection of the countable number of open subsets of S on the right-hand side of (a).

Let $V = \mathbb{Q} \cap [0,1]$, the set of rationals in $[0,1]$. Since V is a countable union of singletons: $V = \{v_n : n \in \mathbb{N}\}$, it is clear that V is $\mathcal{B}[0,1]$-measurable and that $\mathrm{Leb}(V) = 0$. We can include V in an open subset of S of measure at most $4\varepsilon(k)$ as follows:

$$V \subseteq G_k = \bigcup_{n \in \mathbb{N}} [(v_n - \varepsilon(k)2^{-n}, v_n + \varepsilon(k)2^{-n}) \cap S] =: \bigcup_n I_{n,k}.$$

Clearly, $H := \bigcap_k G_k$ satisfies $\mathrm{Leb}(H) = 0$ and $V \subseteq H$. Now, it is a consequence of the *Baire category theorem* (see the appendix to this chapter) that *H is uncountable,* so

(b) *the set H is an uncountable set of measure 0; moreover,*

$$H = \bigcap_k \bigcup_n I_{n,k} \neq \bigcup_n \bigcap_k I_{n,k} = V.$$

Throughout the subject, we have to be careful about interchanging orders of operations.

Chapter 2
Events

2.1. Model for experiment: $(\Omega, \mathcal{F}, \mathbf{P})$

A model for an experiment involving randomness takes the form of a *probability triple* $(\Omega, \mathcal{F}, \mathbf{P})$ in the sense of Section 1.5.

Sample space

Ω is a set called the *sample space*.

Sample point

A point ω of Ω is called a *sample point*.

Event

The σ-algebra $\mathcal{F}$ on Ω is called the family of events, so that an *event* is an element of $\mathcal{F}$, that is, an $\mathcal{F}$-measurable subset of Ω.

By definition of probability triple, $\mathbf{P}$ is a probability measure on $(\Omega, \mathcal{F})$.

2.2. The intuitive meaning

Tyche, Goddess of Chance, chooses a point ω of Ω 'at random' according to the law $\mathbf{P}$ in that, for F in $\mathcal{F}$, $\mathbf{P}(F)$ represents the 'probability' (in the sense understood by our intuition) that the point ω chosen by Tyche belongs to F.

The chosen point ω determines the outcome of the experiment. Thus there is a map

$$\Omega \to \text{set of outcomes},$$
$$\omega \mapsto \text{outcome}.$$

There is no reason why this 'map' (the co-domain lies in our intuition!) should be one-one. Often it is the case that although there is some obvious 'minimal' or 'canonical' model for an experiment, it is better to use some richer model. (For example, we can read off many properties of coin tossing by imbedding the associated random walk in a Brownian motion.)

2.3. Examples of $(\Omega, \mathcal{F})$ pairs

We leave the question of assigning probabilities until later.

(a) *Experiment: Toss coin twice.* We can take

$$\Omega = \{HH, HT, TH, TT\}, \quad \mathcal{F} = \mathcal{P}(\Omega) := \text{set of all subsets of } \Omega.$$

In this model, the intuitive event 'At least one head is obtained' is described by the mathematical event (element of $\mathcal{F}$) $\{HH, HT, TH\}$.

(b) *Experiment: Toss coin infinitely often.* We can take

$$\Omega = \{H, T\}^{\mathbb{N}},$$

so that a typical point ω of Ω is a sequence

$$\omega = (\omega_1, \omega_2, \ldots), \quad \omega_n \in \{H, T\}.$$

We certainly wish to speak of the intuitive event '$\omega_n = W$', where $W \in \{H, T\}$, and it is natural to choose

$$\mathcal{F} = \sigma(\{\omega \in \Omega : \omega_n = W\} : n \in \mathbb{N}, W \in \{H, T\}).$$

Although $\mathcal{F} \neq \mathcal{P}(\Omega)$ (accept this!), it turns out that $\mathcal{F}$ is big enough; for example, we shall see in Section 3.7 that the truth set

$$F = \left\{ \omega : \frac{\#(k \leq n : \omega_k = H)}{n} \to \frac{1}{2} \right\}$$

of the statement

$$\frac{\text{number of heads in } n \text{ tosses}}{n} \to \frac{1}{2}$$

is an element of $\mathcal{F}$.

　　Note that we can use the current model as a more informative model for the experiment in (a), using the map $\omega \mapsto (\omega_1, \omega_2)$ of sample points to outcomes.

(c) *Experiment: Choose a point between 0 and 1 uniformly at random.* Take $\Omega = [0, 1], \mathcal{F} = \mathcal{B}[0, 1], \omega$ signifying the point chosen. In this case, we obviously take $\mathbf{P} = \text{Leb}$. The sense in which this model contains model (b) for the case of a fair coin will be explained later.

2.4. Almost surely (a.s.)

▶A statement S about outcomes is said to be true *almost surely (a.s.)*, or *with probability 1 (w.p.1)*, if

$$F := \{\omega : S(\omega) \text{ is true}\} \in \mathcal{F} \text{ and } \mathbf{P}(F) = 1.$$

(a) **Proposition.** *If $F_n \in \mathcal{F}$ $(n \in \mathbf{N})$ and $\mathbf{P}(F_n) = 1, \forall n$, then*

$$\mathbf{P}(\bigcap_n F_n) = 1.$$

Proof. $\mathbf{P}(F_n^c) = 0, \forall n$, so, by Lemma 1.10(c), $\mathbf{P}(\bigcup_n F_n^c) = 0$. But $\bigcap F_n = (\bigcup F_n^c)^c$. $\qquad\square$

(b) *Something to think about.* Some distinguished philosophers have tried to develop probability without measure theory. One of the reasons for difficulty is the following.

When the discussion (2.3,b) is extended to define the appropriate probability measure for fair coin tossing, the Strong Law of Large Numbers (SLLN) states that $F \in \mathcal{F}$ and $\mathbf{P}(F) = 1$, where F, the truth set of the statement 'proportion of heads in n tosses $\to \frac{1}{2}$', is defined formally in (2.3,b).

Let $\mathcal{A}$ be the set of all maps $\alpha : \mathbf{N} \to \mathbf{N}$ such that $\alpha(1) < \alpha(2) <...$. For $\alpha \in \mathcal{A}$, let

$$F_\alpha = \left\{ \omega : \frac{\#(k \leq n : \omega_{\alpha(k)} = H)}{n} \to \frac{1}{2} \right\}$$

the 'truth set of the Strong Law for the subsequence α'. Then, of course, we have $\mathbf{P}(F_\alpha) = 1, \forall \alpha \in \mathcal{A}$.

Exercise. Prove that

$$\bigcap_{\alpha \in \mathcal{A}} F_\alpha = \emptyset.$$

(*Hint.* For any given ω, find an α ... .)

The moral is that the concept of 'almost surely' gives us (i) absolute precision, but also (ii) enough flexibility to avoid the self-contradictions into which those innocent of measure theory too easily fall. (Of course, since philosophers are pompous where we are precise, they are thought to think deeply ... .)

2.5. Reminder: lim sup, lim inf, ↓ lim, etc.

(a) Let $(x_n : n \in \mathbf{N})$ be a sequence of real numbers. We define

$$\limsup x_n := \inf_m \left\{ \sup_{n \geq m} x_n \right\} = \downarrow \lim_m \left\{ \sup_{n \geq m} x_n \right\} \in [-\infty, \infty].$$

Obviously, $y_m := \sup_{n \geq m} x_n$ is monotone non-increasing in m, so that the limit of the sequence y_m exists in $[-\infty, \infty]$. The use of $\uparrow\lim$ or $\downarrow\lim$ to signify monotone limits will be handy, as will $y_n \downarrow y_\infty$ to signify $y_\infty =\downarrow \lim y_n$.

(b) Analogously,

$$\liminf x_n := \sup_m \left\{ \inf_{n \geq m} x_n \right\} =\uparrow \lim_m \left\{ \inf_{n \geq m} x_n \right\} \in [-\infty, \infty].$$

(c) We have

$$x_n \text{ converges in } [-\infty, \infty] \iff \limsup x_n = \liminf x_n,$$

and then $\lim x_n = \limsup x_n = \liminf x_n$.

▶(d) Note that

(i) if $z > \limsup x_n$, then

$x_n < z$ eventually (that is, for all sufficiently large n)

(ii) if $z < \limsup x_n$, then

$x_n > z$ infinitely often (that is, for infinitely many n).

2.6. Definitions. $\limsup E_n, (E_n, \text{ i.o.})$

The event (in the rigorous formulation: the truth set of the statement)

'number of heads/ number of tosses $\to \frac{1}{2}$'

is built out of simple events such as 'the nth toss results in heads' in a rather complicated way. We need a systematic method of being able to handle complicated combinations of events. The idea of taking lim infs and lim sups of sets provides what is required.

It might be helpful to note the tautology that, if E is an event, then

$$E = \{\omega : \omega \in E\}.$$

Suppose now that $(E_n : n \in \mathbb{N})$ is a sequence of events.

▶(a) We define

$$(E_n, \text{ i.o.}) := (E_n \text{ infinitely often})$$
$$:= \limsup E_n := \bigcap_m \bigcup_{n \geq m} E_n$$
$$= \{\omega : \text{for every } m, \quad \exists n(\omega) \geq m \text{ such that } \omega \in E_{n(\omega)}\}$$
$$= \{\omega : \omega \in E_n \text{ for infinitely many } n\}.$$

▶(b) (**Reverse Fatou Lemma** - *needs FINITENESS of* P)
$$\mathbf{P}(\limsup E_n) \geq \limsup \mathbf{P}(E_n).$$

Proof. Let $G_m := \bigcup_{n \geq m} E_n$. Then (look at the definition in (a)) $G_m \downarrow G$, where $G := \limsup E_n$. By result (1.10,b), $\mathbf{P}(G_m) \downarrow \mathbf{P}(G)$. But, clearly,
$$\mathbf{P}(G_m) \geq \sup_{n \geq m} \mathbf{P}(E_n).$$

Hence,
$$\mathbf{P}(G) \geq \downarrow \lim_m \left\{ \sup_{n \geq m} \mathbf{P}(E_n) \right\} =: \limsup \mathbf{P}(E_n). \qquad \square$$

2.7. First Borel-Cantelli Lemma (BC1)

▶▶   Let $(E_n : n \in \mathbb{N})$ be a sequence of events such that $\sum_n \mathbf{P}(E_n) < \infty$. Then
$$\mathbf{P}(\limsup E_n) = \mathbf{P}(E_n, \text{ i.o.}) = 0.$$

Proof. With the notation of (2.6,b), we have, for each m,
$$\mathbf{P}(G) \leq \mathbf{P}(G_m) \leq \sum_{n \geq m} \mathbf{P}(E_n),$$

using (1.9,b) and (1.10,a). Now let $m \uparrow \infty$. $\square$

Notes. (i) An instructive proof by integration will be given later.

(ii) Many applications of the First Borel-Cantelli Lemma will be given within this course. Interesting applications require concepts of independence, random variables, etc..

2.8. Definitions. $\liminf E_n$, (E_n, ev)

Again suppose that $(E_n : n \in \mathbb{N})$ is a sequence of events.

▶(a) We define
$$(E_n, \text{ ev}) : = (E_n \text{ eventually})$$
$$: = \liminf E_n := \bigcup_m \bigcap_{n \geq m} E_n$$
$$= \{\omega : \text{ for some } m(\omega), \quad \omega \in E_n, \forall n \geq m(\omega)\}$$
$$= \{\omega : \omega \in E_n \text{ for all large } n\}.$$

(b) Note that $(E_n, \text{ ev})^c = (E_n^c, \text{ i.o.})$.

▶▶(c) (**Fatou's Lemma for sets** - *true for ALL measure spaces*)
$$\mathbf{P}(\liminf E_n) \leq \liminf \mathbf{P}(E_n).$$

Exercise. Prove this in analogy with the proof of result (2.6,b), using (1.10,a) rather than (1.10,b).

2.9. Exercise

For an event E, define the indicator function I_E on Ω via

$$\mathrm{I}_E(\omega) := \begin{cases} 1, & \text{if } \omega \in E, \\ 0, & \text{if } \omega \notin E. \end{cases}$$

Let $(E_n : n \in \mathbf{N})$ be a sequence of events. Prove that, for each ω,

$$\mathrm{I}_{\limsup E_n}(\omega) = \limsup \mathrm{I}_{E_n}(\omega),$$

and establish the corresponding result for lim infs.

Chapter 3
Random Variables

Let (S, Σ) be a measurable space, so that Σ is a σ-algebra on S.

3.1. Definitions. Σ-measurable function, $m\Sigma, (m\Sigma)^+, b\Sigma$

Suppose that $h : S \to \mathbf{R}$. For $A \subseteq \mathbf{R}$, define
$$h^{-1}(A) := \{s \in S : h(s) \in A\}$$
Then h is called Σ-*measurable* if $h^{-1} : \mathcal{B} \to \Sigma$, that is, $h^{-1}(A) \in \Sigma, \forall A \in \mathcal{B}$.

So, here is a picture of a Σ-measurable function h:

$$S \xrightarrow{h} \mathbf{R}$$

$$\Sigma \xleftarrow{h^{-1}} \mathcal{B}$$

We write $m\Sigma$ for the class of Σ-measurable functions on S, and $(m\Sigma)^+$ for the class of non-negative elements in $m\Sigma$. We denote by $b\Sigma$ the class of bounded Σ-measurable functions on S.

Note. Because lim sups of sequences even of finite-valued functions may be infinite, and for other reasons, it is convenient to extend these definitions to functions h taking values in $[-\infty, \infty]$ in the obvious way: h is called Σ-*measurable* if $h^{-1} : \mathcal{B}[-\infty, \infty] \to \Sigma$.

Which of the various results stated for real-valued functions extend to functions with values in $[-\infty, \infty]$, and what these extensions are, should be obvious.

Borel function

A function h from a topological space S to $\mathbf{R}$ is called **Borel** if h is $\mathcal{B}(S)$-measurable. The most important case is when S itself is $\mathbf{R}$.

3.2. Elementary Propositions on measurability

(a) *The map h^{-1} preserves all set operations:*
$$h^{-1}(\textstyle\bigcup_\alpha A_\alpha) = \textstyle\bigcup_\alpha h^{-1}(A_\alpha), \quad h^{-1}(A^c) = (h^{-1}(A))^c, \quad \text{etc.}$$
Proof. This is just definition chasing. □

▶(b) *If $C \subseteq B$ and $\sigma(C) = B$, then $h^{-1} : C \to \Sigma \quad \Rightarrow \quad h \in \mathrm{m}\Sigma$.*

Proof. Let $\mathcal{E}$ be the class of elements B in B such that $h^{-1}(B) \in \Sigma$. By result (a), $\mathcal{E}$ is a σ-algebra, and, by hypothesis, $\mathcal{E} \supseteq C$. □

(c) *If S is topological and $h : S \to \mathbf{R}$ is continuous, then h is Borel.*

Proof. Take C to be the class of open subsets of $\mathbf{R}$, and apply result (b). □

▶(d) *For any measurable space (S, Σ), a function $h : S \to \mathbf{R}$ is Σ-measurable if*
$$\{h \le c\} := \{s \in S : h(s) \le c\} \in \Sigma \quad (\forall c \in \mathbf{R}).$$

Proof. Take C to be the class $\pi(\mathbf{R})$ of intervals of the form $(-\infty, c]$, $c \in \mathbf{R}$, and apply result (b). □

Note. Obviously, similar results apply in which $\{h \le c\}$ is replaced by $\{h > c\}$, $\{h \ge c\}$, etc.

3.3. LEMMA. Sums and products of measurable functions are measurable

▶ $\mathrm{m}\Sigma$ *is an algebra over* $\mathbf{R}$, *that is,*

if $\lambda \in \mathbf{R}$ and $h, h_1, h_2 \in \mathrm{m}\Sigma$, then
$$h_1 + h_2 \in \mathrm{m}\Sigma, \quad h_1 h_2 \in \mathrm{m}\Sigma, \quad \lambda h \in \mathrm{m}\Sigma.$$

Example of proof. Let $c \in \mathbf{R}$. Then for $s \in S$, it is clear that $h_1(s) + h_2(s) > c$ if and only if for some rational q, we have
$$h_1(s) > q > c - h_2(s).$$

In other words,
$$\{h_1 + h_2 > c\} = \bigcup_{q \in \mathbf{Q}} (\{h_1 > q\} \cap \{h_2 > c - q\}),$$

a countable union of elements of Σ. □

3.4. Composition Lemma.

If $h \in m\Sigma$ and $f \in m\mathcal{B}$, then $f \circ h \in m\Sigma$.

Proof. Draw the picture:

$$S \xrightarrow{h} \mathbf{R} \xrightarrow{f} \mathbf{R}$$

$$\Sigma \xleftarrow{h^{-1}} \mathcal{B} \xleftarrow{f^{-1}} \mathcal{B}$$

Note. There are obvious generalizations based on the definition (important in more advanced theory): if (S_1, Σ_1) and (S_2, Σ_2) are measurable spaces and $h : S_1 \to S_2$, then h is called Σ_1/Σ_2-measurable if $h^{-1} : \Sigma_2 \to \Sigma_1$. From this point of view, what we have called Σ-*measurable* should read $\Sigma/\mathcal{B}$-*measurable* (or perhaps $\Sigma/\mathcal{B}[-\infty, \infty]$-measurable).

3.5. LEMMA on measurability of infs, lim infs of functions

▶▶   *Let $(h_n : n \in \mathbf{N})$ be a sequence of elements of $m\Sigma$. Then*

(i) $\inf h_n$,   (ii) $\liminf h_n$,   (iii) $\limsup h_n$

are Σ-measurable (into $([-\infty, \infty], \mathcal{B}[-\infty, \infty])$), but we shall still write $\inf h_n \in m\Sigma$ *(for example,). Further,*

(iv) $\{s : \lim h_n(s) \text{ exists in } \mathbf{R}\} \in \Sigma$.

Proof. (i) $\{\inf h_n \geq c\} = \bigcap_n \{h_n \geq c\}$.

(ii) Let $L_n(s) := \inf\{h_r(s) : r \geq n\}$. Then $L_n \in m\Sigma$, by part (i). But

$$L(s) := \liminf h_n(s) = \uparrow \lim L_n(s) = \sup L_n(s),$$

and $\{L \leq c\} = \bigcap_n \{L_n \leq c\} \in \Sigma$.

(iii) This part is now obvious.

(iv) This is also clear because the set on which $\lim h_n$ exists in $\mathbf{R}$ is

$$\{\limsup h_n < \infty\} \cap \{\liminf h_n > -\infty\} \cap g^{-1}(\{0\}),$$

where

$$g := \limsup h_n - \liminf h_n. \qquad \square$$

3.6. Definition. Random variable

▶Let $(\Omega, \mathcal{F})$ be our (sample space, family of events). A *random variable* is an element of $m\mathcal{F}$. Thus,

$$X : \Omega \to \mathbf{R}, \quad X^{-1} : \mathcal{B} \to \mathcal{F}.$$

3.7. Example. Coin tossing

Let $\Omega = \{H, T\}^{\mathbf{N}}$, $\omega = (\omega_1, \omega_2, \ldots)$, $\omega_n \in \{H, T\}$. As in (2.3,b), we define

$$\mathcal{F} = \sigma(\{\omega : \omega_n = W\} : n \in \mathbf{N}, W \in \{H, T\}).$$

Let

$$X_n(\omega) := \begin{cases} 1 & \text{if } \omega_n = H, \\ 0 & \text{if } \omega_n = T. \end{cases}$$

The definition of $\mathcal{F}$ guarantees that each X_n is a random variable. By Lemma 3.3,

$$S_n := X_1 + X_2 + \cdots + X_n = \text{number of heads in } n \text{ tosses}$$

is a random variable.

Next, for $p \in [0, 1]$, we have

$$\Lambda := \left\{ \omega : \frac{\text{number of heads}}{\text{number of tosses}} \to p \right\} = \{\omega : L^+(\omega) = p\} \cap \{\omega : L^-(\omega) = p\},$$

where $L^+ := \limsup n^{-1} S_n$ and L^- is the corresponding lim inf. By Lemma 3.5, $\Lambda \in \mathcal{F}$.

▶▶   Thus, we have taken an important step towards the Strong Law: the result is *meaningful!* It only remains to prove that it is true!

3.8. Definition. σ-algebra generated by a collection of functions on Ω

This is an important idea, discussed further in Section 3.14. (Compare the weakest topology which makes every function in a given family continuous, etc.)

In Example 3.7, we have

a given set Ω,

a family $(X_n : n \in \mathbf{N})$ of maps $X_n : \Omega \to \mathbf{R}$.

The best way to think of the σ-algebra $\mathcal{F}$ in that example is as

$$\mathcal{F} = \sigma(X_n : n \in \mathbf{N})$$

in the sense now to be described.

▶▶Generally, if we have a collection $(Y_\gamma : \gamma \in C)$ of maps $Y_\gamma : \Omega \to \mathbf{R}$, then

$$\mathcal{Y} := \sigma(Y_\gamma : \gamma \in C)$$

is defined to be the smallest σ-algebra $\mathcal{Y}$ on Ω such that each map Y_γ ($\gamma \in C$) is $\mathcal{Y}$-measurable. Clearly,

$$\sigma(Y_\gamma : \gamma \in C) = \sigma(\{\omega \in \Omega : Y_\gamma(\omega) \in B\} : \gamma \in C, B \in \mathcal{B}).$$

If X is a random variable for some $(\Omega, \mathcal{F})$, then, of course, $\sigma(X) \subseteq \mathcal{F}$.

Remarks. (i) The idea introduced in this section is something which you will pick up gradually as you work through the course. Don't *worry* about it now; *think* about it, yes!

(ii) Normally, π-systems come to our aid. For example, if $(X_n : n \subset \mathbf{N})$ is a collection of functions on Ω, and $\mathcal{X}_n$ denotes $\sigma(X_k : k \leq n)$, then the union $\bigcup \mathcal{X}_n$ is a π-system (indeed, an algebra) which generates $\sigma(X_n : n \in \mathbf{N})$.

3.9. Definitions. Law, distribution function

Suppose that X is a random variable carried by some probability triple $(\Omega, \mathcal{F}, \mathbf{P})$. We have

$$\Omega \xrightarrow{X} \mathbf{R}$$

$$[0,1] \xleftarrow{\mathbf{P}} \mathcal{F} \xleftarrow{X^{-1}} \mathcal{B},$$

$$\text{or indeed } [0,1] \xleftarrow{\mathbf{P}} \sigma(X) \xleftarrow{X^{-1}} \mathcal{B}.$$

Define the *law* $\mathcal{L}_X$ of X by

$$\mathcal{L}_X := \mathbf{P} \circ X^{-1}, \qquad \mathcal{L}_X : \mathcal{B} \to [0,1].$$

Then (**Exercise!**) $\mathcal{L}_X$ is a probability measure on $(\mathbf{R}, \mathcal{B})$. Since $\pi(\mathbf{R}) = \{(-\infty, c] : c \in \mathbf{R}\}$ is a π-system which generates $\mathcal{B}$, Uniqueness Lemma 1.6 shows that $\mathcal{L}_X$ is determined by the function $F_X : \mathbf{R} \to [0,1]$ defined as follows:

$$F_X(c) := \mathcal{L}_X(-\infty, c] = \mathbf{P}(X \leq c) = \mathbf{P}\{\omega : X(\omega) \leq c\}.$$

The function F_X is called the *distribution function* of X.

3.10. Properties of distribution functions

Suppose that F is the distribution function $F = F_X$ of some random variable X. Then

(a) $F : \mathbf{R} \to [0,1]$, $F \uparrow$ *(that is, $x \leq y \;\Rightarrow\; F(x) \leq F(y)$),*

(b) $\lim_{x \to \infty} F(x) = 1$, $\lim_{x \to -\infty} F(x) = 0$,

(c)   *F is right-continuous.*

Proof of (c). By using Lemma (1.10,b), we see that

$$\mathbf{P}(X \leq x + n^{-1}) \downarrow \mathbf{P}(X \leq x),$$

and this fact together with the monotonicity of F_X shows that F_X is right-continuous.

Exercise! Clear up any loose ends.

3.11. Existence of random variable with given distribution function

▶If F has the properties (a,b,c) in Section 3.10, then, by analogy with Section 1.8 on the existence of Lebesgue measure, we can construct a unique probability measure $\mathcal{L}$ on $(\mathbf{R}, \mathcal{B})$ such that

$$\mathcal{L}(-\infty, x] = F(x), \forall x.$$

Take $(\Omega, \mathcal{F}, \mathbf{P}) = (\mathbf{R}, \mathcal{B}, \mathcal{L})$, $X(\omega) = \omega$. Then it is tautological that

$$F_X(x) = F(x), \forall x.$$

Note. The measure $\mathcal{L}$ just described is called the *Lebesgue-Stieltjes measure associated with* F. Its existence is proved in the next section.

3.12. Skorokhod representation of a random variable with prescribed distribution function

Again let $F : \mathbf{R} \to [0,1]$ have properties (3.10,a,b,c). We can construct a random variable with distribution function F carried by

$$(\Omega, \mathcal{F}, \mathbf{P}) = ([0,1], \mathcal{B}[0,1], \text{Leb})$$

as follows. Define (the right-hand equalities, which *you* can prove, are there for clarification only)

(a1) $X^+(\omega) := \inf\{z : F(z) > \omega\} = \sup\{y : F(y) \le \omega\},$

(a1) $X^-(\omega) := \inf\{z : F(z) \ge \omega\} = \sup\{y : F(y) < \omega\}.$

The following picture shows cases to watch out for.

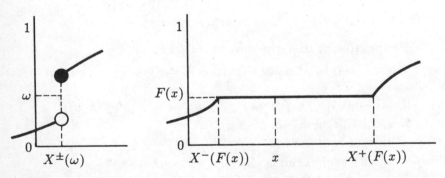

By definition of X^-,

$$(\omega \le F(c)) \quad \Rightarrow \quad (X^-(\omega) \le c).$$

Now,
$$(z > X^-(\omega)) \quad \Rightarrow \quad (F(z) \geq \omega),$$

so, by the right-continuity of F, $F(X^-(\omega)) \geq \omega$, and

$$(X^-(\omega) \leq c) \quad \Rightarrow \quad \left(\omega \leq F(X^-(\omega)) \leq F(c)\right).$$

Thus, $(\omega \leq F(c)) \iff (X^-(\omega) \leq c)$, so that

$$\mathbf{P}(X^- \leq c) = F(c).$$

(b)   *The variable X^- therefore has distribution function F, and the measure $\mathcal{L}$ in Section 3.11 is just the law of X^-.*

It will be important later to know that

(c)   *X^+ also has distribution function F, and that, indeed,*

$$\mathbf{P}(X^+ = X^-) = 1.$$

Proof of (c). By definition of X^+,

$$(\omega < F(c)) \quad \Rightarrow \quad (X^+(\omega) \leq c),$$

so that $F(c) \leq \mathbf{P}(X^+ \leq c)$. Since $X^- \leq X^+$, it is clear that

$$\{X^- \neq X^+\} = \bigcup_{c \in \mathbf{Q}} \{X^- \leq c < X^+\}.$$

But, for every $c \in \mathbf{R}$,

$$\mathbf{P}(X^- \leq c < X^+) = P(\{X^- \leq c\} \backslash \{X^+ \leq c\}) \leq F(c) - F(c) = 0.$$

Since $\mathbf{Q}$ is countable, the result follows. $\square$

Remark. It is in fact true that every experiment you will meet in this (or any other) course can be modelled via the triple $([0,1], \mathcal{B}[0,1], \text{Leb})$. (You will start to be convinced of this by the end of the next chapter.) However, this observation normally has only curiosity value.

3.13. Generated σ-algebras – a discussion

Suppose that $(\Omega, \mathcal{F}, \mathbf{P})$ is a model for some experiment, and that the experiment has been performed, so that (see Section 2.2) Tyche has made her choice of ω.

Let $(Y_\gamma : \gamma \in C)$ be a collection of random variables associated with our experiment, and suppose that someone reports to you the following information about the chosen point ω:

(*) *the values $Y_\gamma(\omega)$, that is, the observed values of the random variables Y_γ ($\gamma \in C$).*

Then the intuitive significance of the σ-algebra $\mathcal{Y} := \sigma(Y_\gamma : \gamma \in C)$ is that it consists precisely of those events F for which, for each and every ω, you can decide whether or not F has occurred (that is, whether or not $\omega \in F$) on the basis of the information (*); the information (*) is precisely equivalent to the following information:

(**) *the values $I_F(\omega)$ ($F \in \mathcal{Y}$).*

(a) **Exercise.** Prove that the σ-algebra $\sigma(Y)$ generated by a single random variable Y is given by

$$\sigma(Y) = Y^{-1}(\mathcal{B}) := (\{\omega : Y(\omega) \in B\} : B \in \mathcal{B}),$$

and that $\sigma(Y)$ is generated by the π-system

$$\pi(Y) := (\{\omega : Y(\omega) \leq x\} : x \in \mathbf{R}) = Y^{-1}(\pi(\mathbf{R})). \qquad \square$$

The following results might help clarify things. Good advice: stop reading this section after (c)! Results (b) and (c) are proved in the appendix to this chapter.

(b) If $Y : \Omega \to \mathbf{R}$, then $Z : \Omega \to \mathbf{R}$ is an $\sigma(Y)$-measurable function if and only if there exists a Borel function $f : \mathbf{R} \to \mathbf{R}$ such that $Z = f(Y)$.

(c) If $Y_1, Y_2, \ldots, Y_n$ are functions from Ω to $\mathbf{R}$, then a function $Z : \Omega \to \mathbf{R}$ is $\sigma(Y_1, Y_2, \ldots, Y_n)$-measurable if and only if there exists a Borel function f on $\mathbf{R}^n$ such that $Z = f(Y_1, Y_2, \ldots, Y_n)$. We shall see in the appendix that the more correct measurability condition on f is that f be '$\mathcal{B}^n$-measurable'.

(d) If $(Y_\gamma : \gamma \in C)$ is a collection (parametrized by the infinite set C) of functions from Ω to $\mathbf{R}$, then $Z : \Omega \to \mathbf{R}$ is $\sigma(Y_\gamma : \gamma \in C)$-measurable if and only if there exists a countable sequence $(\gamma_i : i \in \mathbf{N})$ of elements of C and a Borel function f on $\mathbf{R}^{\mathbf{N}}$ such that

$$Z = f(Y_{\gamma_1}, Y_{\gamma_2}, \ldots).$$

Warning – for the over-enthusiastic only. For uncountable C, $\mathcal{B}(\mathbf{R}^C)$ is much larger than the C-fold product measure space $\prod_{\gamma \in C} \mathcal{B}(\mathbf{R})$. It is the latter rather than the former which gives the appropriate type of f in (d).

3.14. The Monotone-Class Theorem

In the same way that Uniqueness Lemma 1.6 allows us to deduce results about σ-algebras from results about π-systems, the following 'elementary' version of the Monotone-Class Theorem allows us to deduce results about general measurable functions from results about indicators of elements of π-systems. Generally, we shall not use the theorem in the main text, preferring 'just to use bare hands'. However, for product measure in Chapter 8, it becomes indispensable.

THEOREM

►► *Let $\mathcal{H}$ be a class of bounded functions from a set S into $\mathbb{R}$ satisfying the following conditions:*

(i) $\mathcal{H}$ is a vector space over $\mathbb{R}$;

(ii) the constant function 1 is an element of $\mathcal{H}$;

(iii) if (f_n) is a sequence of non-negative functions in $\mathcal{H}$ such that $f_n \uparrow f$ where f is a bounded function on S, then $f \in \mathcal{H}$.

Then if $\mathcal{H}$ contains the indicator function of every set in some π-system $\mathcal{I}$, then $\mathcal{H}$ contains every bounded $\sigma(\mathcal{I})$-measurable function on S.

For proof, see the appendix to this chapter.

Chapter 4
Independence

Let $(\Omega, \mathcal{F}, \mathbf{P})$ be a probability triple.

4.1. Definitions of independence

Note. We focus attention on the σ-algebra formulation (and describe the more familiar forms of independence in terms of it) to acclimatize ourselves to thinking of σ-algebras as the natural means of summarizing information. Section 4.2 shows that the fancy σ-algebra definitions agree with the ones from elementary courses.

Independent σ-algebras

▶Sub-σ-algebras $\mathcal{G}_1, \mathcal{G}_2, \ldots$ of $\mathcal{F}$ are called *independent* if, whenever $G_i \in \mathcal{G}_i$ $(i \in \mathbf{N})$ and $i_1, \ldots, i_n$ are distinct, then

$$\mathbf{P}(G_{i_1} \cap \ldots \cap G_{i_n}) = \prod_{k=1}^{n} \mathbf{P}(G_{i_k}).$$

Independent random variables

▶Random variables $X_1, X_2, \ldots$ are called *independent* if the σ-algebras

$$\sigma(X_1), \sigma(X_2), \ldots$$

are independent.

Independent events

▶Events $E_1, E_2, \ldots$ are called *independent* if the σ-algebras $\mathcal{E}_1, \mathcal{E}_2, \ldots$ are independent, where

$$\mathcal{E}_n \text{ is the } \sigma\text{-algebra } \{\emptyset, E_n, \Omega \backslash E_n, \Omega\}.$$

Since $\mathcal{E}_n = \sigma(I_{E_n})$, it follows that events $E_1, E_2, \ldots$ are independent if and only if the random variables $I_{E_1}, I_{E_2}, \ldots$ are independent.

4.2. The π-system Lemma; and the more familiar definitions

We know from elementary theory that events $E_1, E_2, \ldots$ are independent if and only if whenever $n \in \mathbb{N}$ and $i_1, \ldots, i_n$ are distinct, then

$$P(E_{i_1} \cap \cdots \cap E_{i_n}) = \prod_{k=1}^{n} P(E_{i_k}),$$

corresponding results involving complements of the E_i, etc., being consequences of this.

We now use the Uniqueness Lemma 1.6 to obtain a significant generalization of this idea, **allowing us to study independence via (manageable) π-systems rather than (awkward) σ-algebras.**

Let us concentrate on the case of two σ-algebras.

▶▶(a)   **LEMMA.** *Suppose that $\mathcal{G}$ and $\mathcal{H}$ are sub-σ-algebras of $\mathcal{F}$, and that $\mathcal{I}$ and $\mathcal{J}$ are π-systems with*

$$\sigma(\mathcal{I}) = \mathcal{G}, \quad \sigma(\mathcal{J}) = \mathcal{H}.$$

Then $\mathcal{G}$ and $\mathcal{H}$ are independent if and only if $\mathcal{I}$ and $\mathcal{J}$ are independent in that
$$P(I \cap J) = P(I)P(J), \qquad I \in \mathcal{I}, \ J \in \mathcal{J}.$$

Proof. Suppose that $\mathcal{I}$ and $\mathcal{J}$ are independent. For fixed I in $\mathcal{I}$, the *measures* (check that they *are* measures!)

$$H \mapsto P(I \cap H) \text{ and } H \mapsto P(I)P(H)$$

on $(\Omega, \mathcal{H})$ have the same total mass $P(I)$, and agree on $\mathcal{J}$. By Lemma 1.6, they therefore agree on $\sigma(\mathcal{J}) = \mathcal{H}$. Hence,

$$P(I \cap H) = P(I)P(H), \qquad I \in \mathcal{I}, \ H \in \mathcal{H}.$$

Thus, for fixed H in $\mathcal{H}$, the measures

$$G \mapsto P(G \cap H) \text{ and } G \mapsto P(G)P(H)$$

on $(\Omega, \mathcal{G})$ have the same total mass $P(H)$, and agree on $\mathcal{I}$. They therefore agree on $\sigma(\mathcal{I}) = \mathcal{G}$; and this is what we set out to prove.   □

Suppose now that X and Y are two random variables on $(\Omega, \mathcal{F}, \mathbf{P})$ such that, whenever $x, y \in \mathbf{R}$,

(b) $\qquad\qquad \mathbf{P}(X \leq x; Y \leq y) = \mathbf{P}(X \leq x)\mathbf{P}(Y \leq y).$

Now, (b) says that the π-systems $\pi(X)$ and $\pi(Y)$ (see Section 3.13) are independent. Hence $\sigma(X)$ and $\sigma(Y)$ are independent: that is, X and Y are independent in the sense of Definition 4.1.

In the same way, we can prove that random variables $X_1, X_2, \ldots, X_n$ are independent if and only if

$$\mathbf{P}(X_k \leq x_k : 1 \leq k \leq n) = \prod_{k=1}^{n} \mathbf{P}(X_k \leq x_k),$$

and all the familiar things from elementary theory.

Command: Do Exercise E4.1 now.

4.3. Second Borel-Cantelli Lemma (BC2)

▶▶   *If $(E_n : n \in \mathbf{N})$ is a sequence of* **independent** *events, then*

$$\sum \mathbf{P}(E_n) = \infty \quad \Rightarrow \quad \mathbf{P}(E_n, \text{ i.o.}) = \mathbf{P}(\limsup E_n) = 1.$$

Proof. First, we have

$$(\limsup E_n)^c = \liminf E_n^c = \bigcup_m \bigcap_{n \geq m} E_n^c.$$

With p_n denoting $\mathbf{P}(E_n)$, we have

$$\mathbf{P}\left(\bigcap_{n \geq m} E_n^c\right) = \prod_{n \geq m}(1 - p_n),$$

this equation being true if the condition $\{n \geq m\}$ is replaced by condition $\{r \geq n \geq m\}$, because of independence, and the limit as $r \uparrow \infty$ being justified by the monotonicity of the two sides.

For $x \geq 0$, $\quad 1 - x \leq \exp(-x)$, so that, since $\sum p_n = \infty$,

$$\prod_{n \geq m}(1 - p_n) \leq \exp\left(-\sum_{n \geq m} p_n\right) = 0.$$

So, $\mathbf{P}\left[(\limsup E_n)^c\right] = 0.$ $\qquad\qquad\qquad\qquad\qquad\qquad$ □

Exercise. Prove that if $0 \leq p_n < 1$ and $S := \sum p_n < \infty$, then $\prod(1 - p_n) > 0$. *Hint.* First show that if $S < 1$, then $\prod(1 - p_n) \geq 1 - S$.

4.4. Example

Let $(X_n : n \in \mathbf{N})$ be a sequence of *independent* random variables, each exponentially distributed with rate 1:

$$P(X_n > x) = e^{-x}, \quad x \geq 0.$$

Then, for $\alpha > 0$,

$$P(X_n > \alpha \log n) = n^{-\alpha},$$

so that, using (BC1) and (BC2),

(a0) $\qquad P(X_n > \alpha \log n \text{ for infinitely many } n) = \begin{cases} 0 & \text{if } \alpha > 1, \\ 1 & \text{if } \alpha \leq 1. \end{cases}$

Now let $L := \limsup(X_n / \log n)$. Then

$$P(L \geq 1) \geq P(X_n > \log n, \text{ i.o.}) = 1,$$

and, for $k \in \mathbf{N}$,

$$P(L > 1 + 2k^{-1}) \leq P\left(X_n > (1 + k^{-1}) \log n, \text{ i.o.}\right) = 0.$$

Thus, $\{L > 1\} = \bigcup_k \{L > 1 + 2k^{-1}\}$ is P-null, and hence

$$L = 1 \text{ almost surely.}$$

Something to think about

In the same way, we can prove the finer result

(a1) $\qquad P(X_n > \log n + \alpha \log \log n, \text{ i.o. }) = \begin{cases} 0 & \text{if } \alpha > 1, \\ 1 & \text{if } \alpha \leq 1, \end{cases}$

or, even finer,

(a2) $\quad P(X_n > \log n + \log \log n + \alpha \log \log \log n, \text{ i.o. }) = \begin{cases} 0 & \text{if } \alpha > 1, \\ 1 & \text{if } \alpha \leq 1; \end{cases}$

or etc. By combining in an appropriate way (think about this!) the sequence of statements (a0),(a1),(a2),... with the statement that the union of a countable number of null sets is null while the intersection of a sequence of probability-1 sets has probability 1, we can obviously make remarkably precise statements about the size of the big elements in the sequence (X_n).

I have included in the appendix to this chapter the statement of a truly fantastic theorem about precise description of long-term behaviour: *Strassen's Law.*

A number of exercises in Chapter E are now accessible to you.

4.5. A fundamental question for modelling

Can we construct a sequence $(X_n : n \in \mathbb{N})$ of independent random variables,
X_n having prescribed distribution function F_n? We *have* to be able to answer
Yes to this question – for example, to be able to construct a rigorous model
for the branching-process model of Chapter 0, or indeed for Example 4.4
to make sense. Equation (0.2,b) makes it clear that a Yes answer to our
question is *all* that is needed for a rigorous branching-process model.

The trick answer based on the existence of Lebesgue measure given
in the next section does settle the question. A more satisfying answer is
provided by the theory of product measure, a topic deferred to Chapter 8.

4.6. A coin-tossing model with applications

Let $(\Omega, \mathcal{F}, \mathbb{P})$ be $([0,1], \mathcal{B}[0,1], \text{Leb})$. For $\omega \in \Omega$, expand ω in binary:

$$\omega = 0.\omega_1\omega_2 \dots$$

(The existence of two different expansions of a dyadic rational is not going
to cause any problems because the set $\mathbb{D}$ (say) of dyadic rationals in $[0,1]$
has Lebesgue measure 0 – it is a countable set!) As an **Exercise**, you can
prove that the sequence $(\xi_n : n \in \mathbb{N})$, where

$$\xi_n(\omega) := \omega_n,$$

is a sequence of independent variables each taking the values 0 or 1 with
probability $\frac{1}{2}$ for either possibility. Clearly, $(\xi_n : n \in \mathbb{N})$ provides a model
for coin tossing.

Now define

$$Y_1(\omega) := 0.\omega_1\omega_3\omega_6 \dots ,$$
$$Y_2(\omega) := 0.\omega_2\omega_5\omega_9 \dots ,$$
$$Y_3(\omega) := 0.\omega_4\omega_8\omega_{13} \dots ,$$

and so on. We now need a bit of common sense. Since the sequence

$$\omega_1, \omega_3, \omega_6, \dots$$

has the same 'coin-tossing' properties as the full sequence $(\omega_n : n \in \mathbb{N})$, it
is clear that

Y_1 has the uniform distribution on $[0,1]$;

and similarly for the other Y's.

Since the sequences $(1,3,6,...)$, $(2,5,9,...)$, ... which give rise to $Y_1, Y_2, \ldots$ are disjoint, and therefore correspond to different sets of tosses of our 'coin', it is intuitively obvious that

▶   *$Y_1, Y_2, \ldots$ are independent random variables, each uniformly distributed on $[0, 1]$.*

Now suppose that a sequence $(F_n : n \in \mathsf{N})$ of distribution functions is given. By the Skorokhod representation of Section 3.12, we can find functions g_n on $[0, 1]$ such that

$$X_n := g_n(Y_n) \text{ has distribution function } F_n.$$

But because the Y-variables are independent, the same is obviously true of the X-variables.

▶   *We have therefore succeeded in constructing a family $(X_n : n \in \mathsf{N})$ of independent random variables with prescribed distribution functions.*

Exercise. Satisfy yourself that you could if forced carry through these intuitive arguments rigorously. Obviously, this is again largely a case of utilizing the Uniqueness Lemma 1.6 in much the same way as we did in Section 4.2.

4.7. Notation: IID RVs

Many of the most important problems in probability concern sequences of random variables (RVs) which are *independent and identically distributed* (IID). Thus, if (X_n) is a sequence of IID variables, then the X_n are independent and all have the same distribution function F (say):

$$\mathsf{P}(X_n \leq x) = F(x), \qquad \forall n, \forall x.$$

Of course, we now know that for any given distribution function F, we can construct a triple $(\Omega, \mathcal{F}, \mathsf{P})$ carrying a sequence of IID RVs with common distribution function F. In particular, we can construct a rigorous model for our branching process.

4.8. Stochastic processes; Markov chains

▶A *stochastic process Y parametrized by a set C* is a collection

$$Y = (Y_\gamma : \gamma \in C)$$

of random variables on some triple $(\Omega, \mathcal{F}, \mathsf{P})$. The fundamental question about existence of a stochastic process with prescribed joint distributions is (to all intents and purposes) settled by the famous *Daniell-Kolmogorov theorem*, which is just beyond the scope of this course.

Our concern will be mainly with processes $X = (X_n : n \in \mathbf{Z}^+)$ indexed (or parametrized) by $\mathbf{Z}^+$. We think of X_n as the value of the process X at time n. For $\omega \in \Omega$, the map $n \mapsto X_n(\omega)$ is called the **sample path** of X corresponding to the sample point ω.

A very important example of a stochastic process is provided by a Markov chain.

▶▶Let E be a finite or countable set. Let $P = (p_{ij} : i, j \in E)$ be a *stochastic* $E \times E$ matrix, so that for $i, j \in E$, we have

$$p_{ij} \geq 0, \qquad \sum_k p_{ik} = 1.$$

Let μ be a probability measure on E, so that μ is specified by the values $\mu_i := \mu(\{i\})$, $(i \in E)$. By a *time-homogeneous Markov chain* $Z = (Z_n : n \in \mathbf{Z}^+)$ *on* E *with initial distribution* μ *and 1-step transition matrix* P is meant a stochastic process Z such that, whenever $n \in \mathbf{Z}^+$ and $i_0, i_1, \ldots, i_n \in E$,

(a) $\qquad \mathbf{P}(Z_0 = i_0; Z_1 = i_1; \ldots; Z_n = i_n) = \mu_{i_0} p_{i_0 i_1} \cdots p_{i_{n-1} i_n}.$

Exercise. Give a construction of such a chain Z expressing $Z_n(\omega)$ explicitly in terms of the values at ω of a suitable family of independent random variables. See the appendix to this chapter.

4.9. Monkey typing Shakespeare

Many interesting events must have probability 0 or 1, and we often show that an event F has probability 0 or 1 by using some argument based on independence to show that $\mathbf{P}(F)^2 = \mathbf{P}(F)$.

Here is a silly example, to which we apply a silly method, but one which both illustrates very clearly the use of the monotonicity properties of measures in Lemma 1.10 and has a lot of the flavour of the Kolmogorov 0-1 law. See the 'Easy exercise' towards the end of this section for an instantaneous solution to the problem.

Let us agree that correctly typing WS, the Collected Works of Shakespeare, amounts to typing a particular sequence of N symbols on a typewriter. A monkey types symbols at random, one per unit time, producing an infinite sequence (X_n) of IID RVs with values in the set of all possible symbols. We agree that

$$\varepsilon := \inf\{\mathbf{P}(X_1 = x) : x \text{ is a symbol}\} > 0.$$

Let H be the event that the monkey produces infinitely many copies of WS. Let H_k be the event that the monkey will produce at least k copies of WS in in all, and let $H_{m,k}$ be the event that it will produce at least k copies by time m. Finally, let $H^{(m)}$ be the event that the monkey produces infinitely many copies of WS over the time period $[m+1, \infty)$.

Because the monkey's behaviour over $[1, m]$ is independent of its behaviour over $[m+1, \infty)$, we have

$$\mathbb{P}(H_{m,k} \cap H^{(m)}) = \mathbb{P}(H_{m,k})\mathbb{P}(H^{(m)}).$$

But logic tells us that, for every m, $H^{(m)} = H$! Hence,

$$\mathbb{P}(H_{m,k} \cap H) = \mathbb{P}(H_{m,k})\mathbb{P}(H).$$

But, as $m \uparrow \infty$, $H_{m,k} \uparrow H_k$, and $(H_{m,k} \cap H) \uparrow (H_k \cap H) = H$, it being obvious that $H_k \supseteq H$. Hence, by Lemma 1.10(a),

$$\mathbb{P}(H) = \mathbb{P}(H_k)\mathbb{P}(H).$$

However, as $k \uparrow \infty$, $H_k \downarrow H$, and so, by Lemma 1.10(b),

$$\mathbb{P}(H) = \mathbb{P}(H)\mathbb{P}(H),$$

whence $\mathbb{P}(H) = 0$ or 1.

The Kolmogorov 0-1 law produces a huge class of important events E for which we must have $\mathbb{P}(E) = 0$ or $\mathbb{P}(E) = 1$. Fortunately, it does not tell us which – and it therefore generates a lot of interesting problems!

Easy exercise. Use the Second Borel-Cantelli Lemma to prove that $\mathbb{P}(H) = 1$. *Hint.* Let E_1 be the event that the monkey produces WS right away, that is, during time period $[1, N]$. Then $\mathbb{P}(E_1) \geq \varepsilon^N$.

Tricky exercise (to which we shall return). If the monkey types only capital letters, and is on every occasion equally likely to type any of the 26, how long on average will it take him to produce the sequence

'ABRACADABRA'?

The next three sections involve quite subtle topics which take time to assimilate. They are not strictly necessary for subsequent chapters. The Kolmogorov 0-1 law *is* used in one of our two proofs of the Strong Law for IID RVs, but by that stage a quick martingale proof (of the 0-1 law) will have been provided.

Note. Perhaps the otherwise-wonderful TEX makes its $\mathcal{T}$ too like $\mathcal{I}$. Below, I use $\mathcal{K}$ instead of $\mathcal{I}$ to avoid the confusion. Script X, $\mathcal{X}$, is too like Greek chi, χ, too; but we have to live with that.

4.10. Definition. Tail σ-algebras

▶▶Let $X_1, X_2, \ldots$ be random variables. Define

(a) $$\mathcal{T}_n := \sigma(X_{n+1}, X_{n+2}, \ldots), \quad \mathcal{T} := \bigcap_n \mathcal{T}_n.$$

The σ-algebra $\mathcal{T}$ is called the *tail σ-algebra* of the sequence $(X_n : n \in \mathbb{N})$.

Now, $\mathcal{T}$ contains many important events: for example,

(b1) $$F_1 := (\lim X_k \text{ exists}) := \{\omega : \lim_k X_k(\omega) \text{ exists}\},$$

(b2) $$F_2 := \left(\sum X_k \text{ converges}\right),$$

(b3) $$F_3 := \left(\lim \frac{X_1 + X_2 + \cdots + X_k}{k} \text{ exists}\right).$$

Also, there are many important variables which are in $m\mathcal{T}$: for example,

(c) $$\xi_1 := \limsup \frac{X_1 + X_2 + \cdots + X_k}{k},$$

which may be $\pm\infty$, of course.

Exercise. Prove that F_1, F_2 and F_3 are $\mathcal{T}$-measurable, that the event H in the monkey problem is a tail event, and that the various events of probability 0 and 1 in Section 4.4 are tail events.

Hint – to be read only after you have already tried hard.

Look at F_3 for example. For each n, logic tells us that F_3 is equal to the set

$$F_3^{(n)} := \{\omega : \lim_k \frac{X_{n+1}(\omega) + \cdots + X_{n+k}(\omega)}{k} \text{ exists}\}.$$

Now, $X_{n+1}, X_{n+2}, \ldots$ are all random variables on the triple $(\Omega, \mathcal{T}_n, \mathbb{P})$. That $F_3^{(n)} \in \mathcal{T}_n$ now follows from Lemmas 3.3 and 3.5.

4.11. THEOREM. Kolmogorov's 0-1 Law

▶▶   *Let $(X_n : n \in \mathbb{N})$ be a sequence of **independent** random variables, and let $\mathcal{T}$ be the tail σ-algebra of $(X_n : n \in \mathbb{N})$. Then $\mathcal{T}$ is $\mathbb{P}$-trivial: that is,*
(i) $F \in \mathcal{T} \quad \Rightarrow \quad \mathbb{P}(F) = 0$ or $\mathbb{P}(F) = 1$,
(ii) if ξ is a $\mathcal{T}$-measurable random variable, then, ξ is almost deterministic in that for some constant c in $[-\infty, \infty]$,

$$\mathbb{P}(\xi = c) = 1.$$

We allow $\xi = \pm\infty$ at (ii) for obvious reasons.

Proof of (i). Let

$$\mathcal{X}_n := \sigma(X_1, \dots, X_n), \quad \mathcal{T}_n := \sigma(X_{n+1}, X_{n+2}, \dots).$$

Step 1: We claim that $\mathcal{X}_n$ and $\mathcal{T}_n$ are independent.
Proof of claim. The class $\mathcal{K}$ of events of the form

$$\{\omega : X_i(\omega) \le x_i : 1 \le i \le n\}, \qquad x_i \in \mathbb{R} \cup \{\infty\}$$

is a π-system which generates $\mathcal{X}_n$. The class $\mathcal{J}$ of sets of the form

$$\{\omega : X_j(\omega) \le x_j : n+1 \le j \le n+r\}, \qquad r \in \mathbb{N}, \quad x_j \in \mathbb{R} \cup \{\infty\}$$

is a π-system which generates $\mathcal{T}_n$. But the assumption that the sequence (X_k) is independent implies that $\mathcal{K}$ and $\mathcal{J}$ are independent. Lemma 4.2(a) now clinches our claim.

Step 2: $\mathcal{X}_n$ and $\mathcal{T}$ are independent.
This is obvious because $\mathcal{T} \subseteq \mathcal{T}_n$.

Step 3: We claim that $\mathcal{X}_\infty := \sigma(X_n : n \in \mathbb{N})$ and $\mathcal{T}$ are independent.

Proof of claim. Because $\mathcal{X}_n \subseteq \mathcal{X}_{n+1}$, $\forall n$, the class $\mathcal{K}_\infty := \bigcup \mathcal{X}_n$ is a π-system (it is generally NOT a σ-algebra!) which generates $\mathcal{X}_\infty$. Moreover, $\mathcal{K}_\infty$ and $\mathcal{T}$ are independent, by Step 2. Lemma 4.2(a) again clinches things.

Step 4.
Since $\mathcal{T} \subseteq \mathcal{X}_\infty$, $\mathcal{T}$ is independent of $\mathcal{T}$! Thus,

$$F \in \mathcal{T} \quad \Rightarrow \quad \mathbb{P}(F) = \mathbb{P}(F \cap F) = \mathbb{P}(F)\mathbb{P}(F),$$

and $\mathbb{P}(F) = 0$ or 1. $\qquad\qquad\qquad\qquad\qquad\qquad\qquad\qquad\qquad\qquad\square$

Proof of (ii). By part (i), for every x in $\mathbb{R}$,
$$\mathbb{P}(\xi \le x) = 0 \text{ or } 1.$$
Let $c := \sup\{x : \mathbb{P}(\xi \le x) = 0\}$. Then, if $c = -\infty$, it is clear that $\mathbb{P}(\xi = -\infty) = 1$; and if $c = \infty$, it is clear that $\mathbb{P}(\xi = \infty) = 1$.

So, suppose that c is finite. Then $\mathbb{P}(\xi \le c - 1/n) = 0, \forall n$, so that

$$\mathbb{P}(\bigcup \{\xi \le c - 1/n\}) = \mathbb{P}(\xi < c) = 0,$$

while, since $\mathbb{P}(\xi \le c + 1/n) = 1, \forall n$, we have

$$\mathbb{P}(\bigcap \{\xi \le c + 1/n\}) = \mathbb{P}(\xi \le c) = 1.$$

Hence, $P(\xi = c) = 1$. $\qquad\qquad\qquad\qquad\qquad\qquad\qquad\qquad$ □

Remarks. The examples in Section 4.10 show how striking this result is. For example, *if $X_1, X_2, \ldots$ is a sequence of independent random variables, then*

$$\text{either } \mathbf{P}(\textstyle\sum X_n \text{ converges}) = 0$$

$$\text{or } \mathbf{P}(\textstyle\sum X_n \text{ converges}) = 1.$$

The Three Series Theorem (Theorem 12.5) completely settles the question of which possibility occurs.

So, you can see that the 0-1 law poses numerous interesting questions.

Example. In the branching-process example of Chapter 0, the variable

$$M_\infty := \lim Z_n / \mu^n$$

is measurable on the tail σ-algebra of the sequence $(Z_n : n \in \mathbf{N})$ but need not be almost deterministic. But then the variables $(Z_n : n \in \mathbf{N})$ are not independent.

4.12. Exercise/Warning

Let $Y_0, Y_1, Y_2, \ldots$ be independent random variables with

$$\mathbf{P}(Y_n = +1) = \mathbf{P}(Y_n = -1) = \tfrac{1}{2}, \quad \forall n.$$

For $n \in \mathbf{N}$, define

$$X_n := Y_0 Y_1 \ldots Y_n.$$

Prove that the variables $X_1, X_2, \ldots$ are independent. Define

$$\mathcal{Y} := \sigma(Y_1, Y_2, \ldots), \quad \mathcal{T}_n := \sigma(X_r : r > n).$$

Prove that

$$\mathcal{L} := \bigcap_n \sigma(\mathcal{Y}, \mathcal{T}_n) \neq \sigma\left(\mathcal{Y}, \bigcap_n \mathcal{T}_n\right) =: \mathcal{R}.$$

Hint. Prove that $Y_0 \in m\mathcal{L}$ and that Y_0 is independent of $\mathcal{R}$.

Notes. The phenomenon illustrated by this example tripped up even Kolmogorov and Wiener. The very simple illustration given here was shown to me by Martin Barlow and Ed Perkins. Deciding when we *can* assert that (for $\mathcal{Y}$ a σ-algebra and $(\mathcal{T}_n)$ a decreasing sequence of σ-algebras)

$$\bigcap_n \sigma(\mathcal{Y}, \mathcal{T}_n) = \sigma\left(\mathcal{Y}, \bigcap_n \mathcal{T}_n\right)$$

is a tantalizing problem in many probabilistic contexts.

Chapter 5
Integration

5.0. Notation, etc. $\mu(f) :=: \int f\,d\mu$, $\mu(f; A)$

Let (S, Σ, μ) be a measure space. We are interested in defining for suitable elements f of $m\Sigma$ the (Lebesgue) integral of f with respect to μ, for which we shall use the *alternative notations*:

▶▶
$$\mu(f) :=: \int_S f(s)\mu(ds) :=: \int_S f\,d\mu.$$

It is worth mentioning now that we shall also use the equivalent notations for $A \in \Sigma$:

▶
$$\int_A f(s)\mu(ds) :=: \int_A f\,d\mu :=: \mu(f; A) := \mu(f\mathrm{I}_A)$$

(with a true definition on the extreme right!) It should be clear that, for example,

$$\mu(f; f \geq x) := \mu(f; A), \text{ where } A = \{s \in S : f(s) \geq x\}.$$

Something else worth emphasizing now is that, of course, *summation is a special type of integration.* If $(a_n : n \in \mathbb{N})$ is a sequence of real numbers, then with $S = \mathbb{N}$, $\Sigma = \mathcal{P}(\mathbb{N})$, and μ the measure on (S, Σ) with $\mu(\{k\}) = 1$ for every k in $\mathbb{N}$, then $s \mapsto a_s$ is μ-integrable if and only if $\sum |a_n| < \infty$, and then

$$\sum a_n = \int_S a_s\mu(ds) = \int_S a\,d\mu.$$

We begin by considering the integral of a function f in $(m\Sigma)^+$, *allowing such an f to take values in the extended half-line* $[0, \infty]$.

5.1. Integrals of non-negative simple functions, SF^+

If A is an element of Σ, we define

$$\mu_0(I_A) := \mu(A) \leq \infty.$$

The use of μ_0 rather than μ signifies that we currently have only a naive integral defined for simple functions.

An element f of $(m\Sigma)^+$ is called *simple*, and we shall then write $f \in SF^+$, if f may be written as a finite sum

(a) $$f = \sum_{k=1}^{m} a_k I_{A_k}$$

where $a_k \in [0, \infty]$ and $A_k \in \Sigma$. We then define

(b) $$\mu_0(f) = \sum a_k \mu(A_k) \leq \infty \qquad \text{(with } 0.\infty := 0 =: \infty.0\text{)}.$$

The first point to be checked is that $\mu_0(f)$ is well-defined; for f will have many different representations of the form (a), and we must ensure that they yield the same value of $\mu_0(f)$ in (b). Various desirable properties also need to be checked, namely (c), (d) and (e) now to be stated:

(c)   if $f, g \in SF^+$ and $\mu(f \neq g) = 0$ then $\mu_0(f) = \mu_0(g)$;

(d)   ('Linearity') if $f, g \in SF^+$ and $c \geq 0$ then $f + g$ and cf are in SF^+, and

$$\mu_0(f + g) = \mu_0(f) + \mu_0(g), \quad \mu_0(cf) = c\mu_0(f);$$

(e)   (Monotonicity) if $f, g \in SF^+$ and $f \leq g$, then $\mu_0(f) \leq \mu_0(g)$;

(f)   if $f, g \in SF^+$ then $f \wedge g$ and $f \vee g$ are in SF^+.

Checking all the properties just mentioned is a little messy, but it involves no point of substance, and in particular no *analysis*. We skip this, and turn our attention to what matters: the Monotone-Convergence Theorem.

5.2. Definition of $\mu(f)$, $f \in (m\Sigma)^+$

▶For $f \in (m\Sigma)^+$ we define

(a) $$\mu(f) := \sup\{\mu_0(h) : h \in SF^+, h \leq f\} \leq \infty.$$

Clearly, for $f \in SF^+$, we have $\mu(f) = \mu_0(f)$.

The following result is important.

LEMMA

▶(b)   If $f \in (\mathrm{m}\Sigma)^+$ and $\mu(f) = 0$, then

$$\mu(\{f > 0\}) = 0.$$

Proof. Obviously, $\{f > 0\} = \uparrow \lim\{f > n^{-1}\}$. Hence, using (1.10,a), we see that if $\mu(\{f > 0\}) > 0$, then, for some n, $\mu(\{f > n^{-1}\}) > 0$, and then

$$\mu(f) \geq \mu_0(n^{-1}\mathrm{I}_{\{f > 1/n\}}) > 0. \qquad \square$$

5.3. Monotone-Convergence Theorem (MON)

▶▶(a)   **If (f_n) is a sequence of elements of $(\mathrm{m}\Sigma)^+$ such that $f_n \uparrow f$, then**

$$\mu(f_n) \uparrow \mu(f) \leq \infty,$$

or, in other notation,

$$\int_S f_n(s)\mu(ds) \uparrow \int_S f(s)\mu(ds).$$

This theorem is really all there is to integration theory. We shall see that other key results such as the Fatou Lemma and the Dominated-Convergence Theorem follow trivially from it.

The (MON) theorem is proved in the Appendix. Obviously, the theorem relates very closely to Lemma 1.10(a), the monotonicity result for measures. The proof of (MON) is not at all difficult, and may be read once you have looked at the following definition of $\alpha^{(r)}$.

It is convenient to have an explicit way given $f \in (\mathrm{m}\Sigma)^+$ of obtaining a sequence $f^{(r)}$ of simple functions such that $f^{(r)} \uparrow f$. For $r \in \mathbf{N}$, define the rth **staircase function** $\alpha^{(r)} : [0,\infty] \to [0,\infty]$ as follows:

(b) $\alpha^{(r)}(x) := \begin{cases} 0 & \text{if } x = 0, \\ (i-1)2^{-r} & \text{if } (i-1)2^{-r} < x \leq i2^{-r} \leq r \quad (i \in \mathbf{N}), \\ r & \text{if } x > r. \end{cases}$

Then $f^{(r)} = \alpha^{(r)} \circ f$ satisfies $f^{(r)} \in SF^+$, and $f^{(r)} \uparrow f$ so that, by (MON),

$$\mu(f) = \uparrow \lim \mu(f^{(r)}) = \uparrow \lim \mu_0(f^{(r)}).$$

We have made $\alpha^{(r)}$ *left-continuous* so that if $f_n \uparrow f$ then $\alpha^{(r)}(f_n) \uparrow \alpha^{(r)}(f)$.

Often, we need to apply convergence theorems such as (MON) where the hypothesis ($f_n \uparrow f$ in the case of (MON)) holds almost everywhere rather than everywhere. Let us see how such adjustments may be made.

(c)   *If $f, g \in (m\Sigma)^+$ and $f = g$ (a.e.), then $\mu(f) = \mu(g)$.*

Proof. Let $f^{(r)} = \alpha^{(r)} \circ f$, $g^{(r)} = \alpha^{(r)} \circ g$. Then $f^{(r)} = g^{(r)}$ (a.e.) and so, by (5.1,c), $\mu(f^{(r)}) = \mu(g^{(r)})$. Now let $r \uparrow \infty$, and use (MON).   □

▶(d)   *If $f \in (m\Sigma)^+$ and (f_n) is a sequence in $(m\Sigma)^+$ such that, except on a μ-null set N, $f_n \uparrow f$. Then*

$$\mu(f_n) \uparrow \mu(f).$$

Proof. We have $\mu(f) = \mu(fI_{S\setminus N})$ and $\mu(f_n) = \mu(f_n I_{S\setminus N})$. But $f_n I_{S\setminus N} \uparrow fI_{S\setminus N}$ *everywhere.* The result now follows from (MON).   □

From now on, (MON) is understood to include this extension. We do not bother to spell out such extensions for the other convergence theorems, often stating results with 'almost everywhere' but proving them under the assumption that the exceptional null set is empty.

Note on the Riemann integral

If, for example, f is a non-negative Riemann integrable function on $([0, 1], \mathcal{B}[0, 1], \text{Leb})$ with Riemann integral I, then there exists an increasing sequence (L_n) of elements of SF^+ and a decreasing sequence (U_n) of elements of SF^+ such that

$$L_n \uparrow L \leq f, \quad U_n \downarrow U \geq f$$

and $\mu(L_n) \uparrow \text{I}$, $\mu(U_n) \downarrow \text{I}$. If we define

$$\tilde{f} = \begin{cases} L & \text{if } L = U, \\ 0 & \text{otherwise,} \end{cases}$$

then it is clear that $\tilde{f}$ is Borel measurable, while (since $\mu(L) = \mu(U) = 1$) $\{f \neq \tilde{f}\}$ is a subset of the Borel set $\{L \neq U\}$ which Lemma 5.2(b) shows to be of measure 0. So f is Lebesgue measurable (see Section A1.11) and the Riemann integral of f equals the integral of f associated with $([0, 1], Leb[0, 1], \text{Leb})$, $Leb[0, 1]$ denoting the σ-algebra of Lebesgue measurable subsets of $[0,1]$.

5.4. The Fatou Lemmas for functions

▶▶(a)   **(FATOU)** *For a sequence (f_n) in $(m\Sigma)^+$,*

$$\mu(\liminf f_n) \leq \liminf \mu(f_n).$$

Proof. We have

$$(*) \qquad \liminf_n f_n = \uparrow \lim g_k, \quad \text{where } g_k := \inf_{n \geq k} f_n.$$

For $n \geq k$, we have $f_n \geq g_k$, so that $\mu(f_n) \geq \mu(g_k)$, whence

$$\mu(g_k) \leq \inf_{n \geq k} \mu(f_n);$$

and on combining this with an application of (MON) to ($*$), we obtain

$$\mu(\liminf_n f_n) = \uparrow \lim_k \mu(g_k) \leq \uparrow \lim_k \inf_{n \geq k} \mu(f_n)$$

$$=: \liminf_n \mu(f_n). \qquad \square$$

Reverse Fatou Lemma

▶(b)   *If (f_n) is a sequence in $(m\Sigma)^+$ such that for some g in $(m\Sigma)^+$, we have $f_n \leq g, \forall n$, and $\mu(g) < \infty$, then*

$$\mu(\limsup f_n) \geq \limsup \mu(f_n).$$

Proof. Apply (FATOU) to the sequence $(g - f_n)$. $\qquad \square$

5.5. 'Linearity'

For $\alpha, \beta \in \mathbf{R}^+$ and $f, g \in (m\Sigma)^+$,

$$\mu(\alpha f + \beta g) = \alpha\mu(f) + \beta\mu(g) \quad (\leq \infty).$$

Proof. Approximate f and g from below by simple functions, apply (5.1,d) to the simple functions, and then use (MON). $\qquad \square$

5.6. Positive and negative parts of f

For $f \in m\Sigma$, we write $f = f^+ - f^-$, where

$$f^+(s) := \max(f(s), 0), \quad f^-(s) := \max(-f(s), 0).$$

Then $f^+, f^- \in (m\Sigma)^+$, and $|f| = f^+ + f^-$.

5.7. Integrable function, $\mathcal{L}^1(S, \Sigma, \mu)$

▶For $f \in m\Sigma$, we say that f is *μ-integrable*, and write

$$f \in \mathcal{L}^1(S, \Sigma, \mu)$$

if

$$\mu(|f|) = \mu(f^+) + \mu(f^-) < \infty,$$

and then we define

$$\int f d\mu := \mu(f) := \mu(f^+) - \mu(f^-).$$

Note that, for $f \in \mathcal{L}^1(S, \Sigma, \mu)$,

▶ $$|\mu(f)| \leq \mu(|f|),$$

the familiar rule that *the modulus of the integral is less than or equal to the integral of the modulus.*

We write $\mathcal{L}^1(S, \Sigma, \mu)^+$ for the class of non-negative elements in $\mathcal{L}^1(S, \Sigma, \mu)$.

5.8. Linearity

For $\alpha, \beta \in \mathbf{R}$ and $f, g \in \mathcal{L}^1(S, \Sigma, \mu)$,

$$\alpha f + \beta g \in \mathcal{L}^1(S, \Sigma, \mu)$$

and

$$\mu(\alpha f + \beta g) = \alpha \mu(f) + \beta \mu(g).$$

Proof. This is a totally routine consequence of the result in Section 5.5. □

5.9. Dominated-Convergence Theorem (DOM)

▶   *Suppose that $f_n, f \in m\Sigma$, that $f_n(s) \to f(s)$ for every s in S and that the sequence (f_n) is **dominated** by an element g of $\mathcal{L}^1(S, \Sigma, \mu)^+$:*

$$|f_n(s)| \leq g(s), \quad \forall s \in S, \forall n \in \mathbf{N},$$

where $\mu(g) < \infty$. Then

$$f_n \to f \text{ in } \mathcal{L}^1(S, \Sigma, \mu): \text{ that is, } \mu(|f_n - f|) \to 0,$$

whence

$$\mu(f_n) \to \mu(f).$$

Command: Do Exercise E5.1 now.

Proof. We have $|f_n - f| \le 2g$, where $\mu(2g) < \infty$, so by the reverse Fatou Lemma 5.4(b),

$$\limsup \mu(|f_n - f|) \le \mu(\limsup |f_n - f|) = \mu(0) = 0.$$

Since

$$|\mu(f_n) - \mu(f)| = |\mu(f_n - f)| \le \mu(|f_n - f|),$$

the theorem is proved. $\square$

5.10. Scheffé's Lemma (SCHEFFÉ)

►(i)   *Suppose that $f_n, f \in \mathcal{L}^1(S, \Sigma, \mu)^+$; in particular, f_n and f are non-negative. Suppose that $f_n \to f$ (a.e.). Then*

$$\mu(|f_n - f|) \to 0 \text{ if and only if } \mu(f_n) \to \mu(f).$$

Proof. The 'only if' part is trivial.

Suppose now that

(a) $\mu(f_n) \to \mu(f).$

Since $(f_n - f)^- \le f$, (DOM) shows that

(b) $\mu((f_n - f)^-) \to 0.$

Next,

$$\mu((f_n - f)^+) = \mu(f_n - f; f_n \ge f)$$
$$= \mu(f_n) - \mu(f) - \mu(f_n - f; f_n < f).$$

But

$$|\mu(f_n - f; f_n < f)| \le |\mu((f_n - f)^-)| \to 0$$

so that (a) and (b) together imply that

(c) $\mu((f_n - f)^+) \to 0.$

Of course, (b) and (c) now yield the desired result. $\square$

Here is the second part of Scheffé's Lemma.

►(ii)   *Suppose that $f_n, f \in \mathcal{L}^1(S, \Sigma, \mu)$ and that $f_n \to f$ (a.e.). Then*

$$\mu(|f_n - f|) \to 0 \text{ if and only if } \mu(|f_n|) \to \mu(|f|).$$

Exercise. Prove the 'if' part of (ii) by using Fatou's Lemma to show that $\mu(f_n^\pm) \to \mu(f^\pm)$, and then applying (i). Of course, the 'only if' part is trivial.

5.11. Remark on uniform integrability

The theory of uniform integrability, which we shall establish later for probability triples, gives better insight into the matter of convergence of integrals.

5.12. The standard machine

What I call the standard machine is a much cruder alternative to the Monotone-Class Theorem.

The idea is that to prove that a 'linear' result is true for all functions h in a space such as $\mathcal{L}^1(S, \Sigma, \mu)$,

- first, we show the result is true for the case when h is an indicator function – which it normally is by definition;

- then, we use linearity to obtain the result for h in SF$^+$;

- next, we use (MON) to obtain the result for $h \in (m\Sigma)^+$, integrability conditions on h usually being superfluous at this stage;

- finally, we show, by writing $h = h^+ - h^-$ and using linearity, that the claimed result is true.

It seems to me that, when it works, it is easier to 'watch the standard machine work' than to appeal to the monotone-class result, though there are times when the greater subtlety of the Monotone-Class Theorem is essential.

5.13. Integrals over subsets

Recall that for $f \in (m\Sigma)^+$, we set, for $A \in \Sigma$,

$$\int_A f d\mu =: \mu(f; A) := \mu(fI_A).$$

If we really want to integrate f over A, we should integrate the restriction $f|_A$ with respect to the measure μ_A (say) which is μ restricted to the measure space (A, Σ_A), Σ_A denoting the σ-algebra of subsets of A which belong to Σ. So we ought to prove that

(a) $$\mu_A(f|_A) = \mu(f; A).$$

The standard machine does this. If f is the indicator of a set B in A, then both sides of (a) are just $\mu(A \cap B)$; etc. We discover that *for $f \in m\Sigma$, we have $f|_A \in m\Sigma_A$; and then*

$$f|_A \in \mathcal{L}^1(A, \Sigma_A, \mu_A) \text{ if and only if } fI_A \in \mathcal{L}^1(S, \Sigma, \mu),$$

in which case (a) *holds.*

5.14. The measure $f\mu$, $f \in (m\Sigma)^+$

Let $f \in (m\Sigma)^+$. For $A \in \Sigma$, define

(a) $\qquad (f\mu)(A) := \mu(f;A) := \mu(fI_A).$

A trivial **Exercise** on the results of Section 5.5 and (MON) shows that

(b) $\qquad (f\mu)$ *is a measure on* (S,Σ).

For $h \in (m\Sigma)^+$, and $A \in \Sigma$, we can conjecture that

(c) $\qquad (h(f\mu))(A) := (f\mu)(hI_A) = \mu(fhI_A).$

If h is the indicator of a set in Σ, then (c) is immediate by definition. Our standard machine produces (c), so that we have

(d) $\qquad h(f\mu) = (hf)\mu.$

Result (d) is often used in the following form:

▶(e) *if* $f \in (m\Sigma)^+$ *and* $h \in (m\Sigma)$, *then* $h \in \mathcal{L}^1(S,\Sigma,f\mu)$ *if and only if* $fh \in \mathcal{L}^1(S,\Sigma,\mu)$ *and then* $(f\mu)(h) = \mu(fh)$.

Proof. We need only prove this for $h \geq 0$ in which case it merely says that the measures at (d) agree on S. $\qquad\square$

Terminology, and the Radon-Nikodým theorem

If λ denotes the measure $f\mu$ on (S,Σ), we say that λ has *density* f relative to μ, and express this in symbols via

$$d\lambda/d\mu = f.$$

We note that in this case, we have for $F \in \Sigma$:

(f) $\qquad \mu(F) = 0$ implies that $\lambda(F) = 0;$

so that only certain measures have density relative to μ. The **Radon-Nikodým theorem** (proved in Chapter 14) tells us that

(g) *if* μ *and* λ *are* σ*-finite measures on* (S,Σ) *such that* (f) *holds, then* $\lambda = f\mu$ *for some* $f \in (m\Sigma)^+$.

Chapter 6
Expectation

6.0. Introductory remarks

We work with a **probability triple** $(\Omega, \mathcal{F}, \mathbf{P})$, *and write* $\mathcal{L}^r$ *for* $\mathcal{L}^r(\Omega, \mathcal{F}, \mathbf{P})$. Recall that a *random variable* (RV) is an element of $m\mathcal{F}$, that is an $\mathcal{F}$-measurable function from Ω to $\mathbf{R}$.

Expectation is just the integral relative to $\mathbf{P}$.

Jensen's inequality, which makes critical use of the fact that $\mathbf{P}(\Omega) = 1$, is very useful and powerful: it implies the Schwarz, Hölder, ... inequalities for general (S, Σ, μ). (See Section 6.13.)

We study the *geometry of the space* $\mathcal{L}^2(\Omega, \mathcal{F}, \mathbf{P})$ in some detail, with a view to several later applications.

6.1. Definition of expectation

For a random variable $X \in \mathcal{L}^1 = \mathcal{L}^1(\Omega, \mathcal{F}, \mathbf{P})$, we define the *expectation* $\mathbf{E}(X)$ *of* X by

$$\mathbf{E}(X) := \int_\Omega X dP = \int_\Omega X(\omega)\mathbf{P}(d\omega).$$

We also define $\mathbf{E}(X)$ $(\leq \infty)$ for $X \in (m\mathcal{F})^+$. In short, $\mathbf{E}(X) = \mathbf{P}(X)$.

That our present definitions agree with those in terms of probability density function (if it exists) etc. will be confirmed in Section 6.12.

6.2. Convergence theorems

Suppose that (X_n) *is a sequence of RVs, that* X *is a RV, and that* $X_n \to X$ *almost surely:*

$$\mathbf{P}(X_n \to X) = 1.$$

We rephrase the convergence theorems of Chapter 5 in our new notation:

▶▶(MON) if $0 \leq X_n \uparrow X$, *then* $E(X_n) \uparrow E(X) \leq \infty$;

▶▶(FATOU) *if* $X_n \geq 0$, *then* $E(X) \leq \liminf E(X_n)$;

▶(DOM) *if* $|X_n(\omega)| \leq Y(\omega)$ $\forall(n,\omega)$, *where* $E(Y) < \infty$, *then*

$$E(|X_n - X|) \to 0,$$

so that
$$E(X_n) \to E(X);$$

▶(SCHEFFÉ) *if* $E(|X_n|) \to E(|X|)$, *then*

$$E(|X_n - X|) \to 0;$$

▶▶(BDD) *if for some finite constant* K, $|X_n(\omega)| \leq K, \forall(n,\omega)$, *then*

$$E(|X_n - X|) \to 0.$$

The newly-added *Bounded Convergence Theorem* (BDD) is an immediate consequence of (DOM), obtained by taking $Y(\omega) = K$, $\forall\omega$; because of the fact that $P(\Omega) = 1$, we have $E(Y) < \infty$. It has a direct elementary proof which we shall examine in Section 13.7; but you might well be able to provide it now.

As has been mentioned previously, *uniform integrability* is the key concept which gives a proper understanding of convergence theorems. We shall study this, via the elementary (BDD) result, in Chapter 13.

6.3. The notation $E(X; F)$

For $X \in \mathcal{L}^1$ (or $(m\mathcal{F})^+$) and $F \in \mathcal{F}$, we define

▶ $$E(X; F) := \int_F X(\omega) P(d\omega) := E(XI_F),$$

where, as ever,
$$I_F(\omega) := \begin{cases} 1 & \text{if } \omega \in F, \\ 0 & \text{if } \omega \notin F. \end{cases}$$

Of course, this tallies with the $\mu(f; A)$ notation of Chapter 5.

6.4. Markov's inequality

Suppose that $Z \in m\mathcal{F}$ *and that* $g : \mathbf{R} \to [0, \infty]$ *is* $\mathcal{B}$-*measurable and non-decreasing. (We know that* $g(Z) = g \circ Z \in (m\mathcal{F})^+$.) *Then*

▶ $$Eg(Z) \geq E(g(Z); Z \geq c) \geq g(c)P(Z \geq c).$$

Examples: for $Z \in (m\mathcal{F})^+$, $cP(Z \geq c) \leq E(Z)$, $(c > 0)$,

for $X \in \mathcal{L}^1$, $cP(|X| \geq c) \leq E(|X|)$ $(c > 0)$.

▶▶*Considerable strength can often be obtained by choosing the optimum* θ *for* c *in*

▶ $P(Y > c) \leq e^{-\theta c} E(e^{\theta Y})$, $(\theta > 0,$ $c \in \mathbf{R})$.

6.5. Sums of non-negative RVs

We collect together some useful results.

(a) *If* $X \in (m\mathcal{F})^+$ *and* $E(X) < \infty$, *then* $P(X < \infty) = 1$. This is obvious.

▶(b) *If* (Z_k) *is a sequence in* $(m\mathcal{F})^+$, *then*

$$E(\sum Z_k) = \sum E(Z_k) \leq \infty.$$

This is an obvious consequence of linearity and (MON).

▶(c) *If* (Z_k) *is a sequence in* $(m\mathcal{F})^+$ *such that* $\sum E(Z_k) < \infty$, *then*

$$\sum Z_k < \infty \text{ (a.s.) and so } Z_k \to 0 \text{ (a.s.)}$$

This is an immediate consequence of (a) and (b).

(d) *The First Borel-Cantelli Lemma is a consequence of* (c). For suppose that (F_k) is a sequence of events such that $\sum P(F_k) < \infty$. Take $Z_k = I_{F_k}$. Then $E(Z_k) = P(F_k)$ and, by (c),

$$\sum I_{F_k} = \text{number of events } F_k \text{ which occur}$$

is a.s. finite.

6.6. Jensen's inequality for convex functions

▶▶A function $c : G \to \mathbf{R}$, where G is an open subinterval of $\mathbf{R}$, is called **convex** on G if its graph lies below any of its chords: for $x, y \in G$ and $0 \leq p = 1 - q \leq 1$,
$$c(px + qy) \leq pc(x) + qc(y).$$

It will be explained below that c is automatically continuous on G. If c is twice-differentiable on G, then c is convex if and only if $c'' \geq 0$.

▶*Important examples of convex functions:* $|x|, x^2, e^{\theta x}(\theta \in \mathbf{R})$.

THEOREM. Jensen's inequality

▶▶   *Suppose that $c : G \to \mathbf{R}$ is a convex function on an open subinterval G of $\mathbf{R}$ and that X is a random variable such that*

$$E(|X|) < \infty, \qquad P(X \in G) = 1, \qquad E|c(X)| < \infty.$$

Then
$$Ec(X) \geq c(E(X)).$$

Proof. The fact that c is convex may be rewritten as follows: for $u, v, w \in G$ with $u < v < w$, we have

$$\Delta_{u,v} \leq \Delta_{v,w}, \text{ where } \Delta_{u,v} := \frac{c(v) - c(u)}{v - u}.$$

It is now clear (why?!) that c is continuous on G, and that for each v in G the monotone limits

$$(D_- c)(v) := \uparrow \lim_{u \uparrow v} \Delta_{u,v}, \qquad (D_+ c)(v) := \downarrow \lim_{w \downarrow v} \Delta_{v,w}$$

exist and satisfy $(D_- c)(v) \leq (D_+ c)(v)$. The functions $D_- c$ and $D_+ c$ are non-decreasing, and for every v in G, for any m in $[(D_- c)(v), (D_+ c)(v)]$ we have

$$c(x) \geq m(x - v) + c(v), \qquad x \in G.$$

In particular, we have, almost surely, for $\mu := E(X)$,

$$c(X) \geq m(X - \mu) + c(\mu), \quad m \in [(D_- c)(\mu), (D_+ c)(\mu)]$$

and Jensen's inequality follows on taking expectations.   □

Remark. For later use, we shall need the obvious fact that

(a) $c(x) = \sup_{q \in G}[(D_- c)(q)(x - q) + c(q)] = \sup_n (a_n x + b_n) \qquad (x \in G)$

for some sequences (a_n) and (b_n) in $\mathbf{R}$. (Recall that c is continuous.)

6.7. Monotonicity of $\mathcal{L}^p$ norms

▶▶For $1 \leq p < \infty$, we say that $X \in \mathcal{L}^p = \mathcal{L}^p(\Omega, \mathcal{F}, P)$ if

$$E(|X|^p) < \infty,$$

and then we define

▶▶
$$\|X\|_p := \{\mathbb{E}(|X|^p)\}^{\frac{1}{p}}.$$

The monotonicity property referred to in the section title is the following:

▶(a) *if $1 \le p \le r < \infty$ and $Y \in \mathcal{L}^r$, then $Y \in \mathcal{L}^p$ and*

$$\|Y\|_p \le \|Y\|_r.$$

▶*Proof.* For $n \in \mathbb{N}$, define

$$X_n(\omega) := \{|Y(\omega)| \wedge n\}^p.$$

Then X_n is bounded so that X_n and $X_n^{r/p}$ are both in $\mathcal{L}^1$. Taking $c(x) = x^{r/p}$ on $(0, \infty)$, we conclude from Jensen's inequality that

$$(\mathbb{E}X_n)^{r/p} \le \mathbb{E}(X_n^{r/p}) = \mathbb{E}[(|Y| \wedge n)^r] \le \mathbb{E}(|Y|^r).$$

Now let $n \uparrow \infty$ and use (MON) to obtain the desired result. □

Note. The proof is marked with a ▶ because it illustrates a simple but effective use of truncation.

'Vector-space property' of $\mathcal{L}^p$

(b)   Since, for $a, b \in \mathbb{R}^+$, we have

$$(a + b)^p \le [2\max(a, b)]^p \le 2^p(a^p + b^p),$$

$\mathcal{L}^p$ is *'obviously a vector space'* except that difficulties with infinities can destroy the associative law. Because a function in $\mathcal{L}^p$ can be infinite only on a null set, this is not a serious annoyance. See the discussion of quotienting at the end of Section 6.9.

6.8. The Schwarz inequality

▶▶(a) *If X and Y are in $\mathcal{L}^2$, then $XY \in \mathcal{L}^1$, and*

$$|\mathbb{E}(XY)| \le \mathbb{E}(|XY|) \le \|X\|_2 \|Y\|_2.$$

Remark. You will have seen many versions of this result and of its proof before. We use truncation to make the argument rigorous.

Proof. By considering $|X|$ and $|Y|$ instead of X and Y, we can and do restrict attention to the case when $X \ge 0$, $Y \ge 0$.

Write $X_n := X \wedge n$, $Y_n := Y \wedge n$, so that X_n and Y_n are bounded. For any $a, b \in \mathbf{R}$,

$$0 \leq \mathsf{E}[(aX_n + bY_n)^2]$$
$$= a^2 \mathsf{E}(X_n^2) + 2ab\mathsf{E}(X_n Y_n) + b^2 \mathsf{E}(Y_n^2),$$

and since the quadratic in a/b (or b/a, or...) does not have two distinct real roots,

$$\{2\mathsf{E}(X_n Y_n)\}^2 \leq 4\mathsf{E}(X_n^2)\mathsf{E}(Y_n^2) \leq 4\mathsf{E}(X^2)\mathsf{E}(Y^2).$$

Now let $n \uparrow \infty$ using (MON).   □

The following is an immediate consequence of (a):

(b) *if X and Y are in $\mathcal{L}^2$, then so is $X + Y$, and we have the triangle law:*

$$\|X + Y\|_2 \leq \|X\|_2 + \|Y\|_2.$$

Remark. The Schwarz inequality is true for *any* measure space – see Section 6.13, which gives the extensions of (a) and (b) to $\mathcal{L}^p$.

6.9. $\mathcal{L}^2$: Pythagoras, covariance, etc.

In this section, we take a brief look at the geometry of $\mathcal{L}^2$ and at its connections with probabilistic concepts such as covariance, correlation, etc.

Covariance and variance

If $X, Y \in \mathcal{L}^2$, then by the monotonicity of norms, $X, Y \in \mathcal{L}^1$, so that we may define

$$\mu_X := \mathsf{E}(X), \quad \mu_Y := \mathsf{E}(Y).$$

Since the constant functions with values μ_X, μ_Y are in $\mathcal{L}^2$, we see that

(a) $$\tilde{X} := X - \mu_X, \quad \tilde{Y} := Y - \mu_Y$$

are in $\mathcal{L}^2$. By the Schwarz inequality, $\tilde{X}\tilde{Y} \in \mathcal{L}^1$, and so we may define

(b) $$\mathrm{Cov}(X, Y) := \mathsf{E}(\tilde{X}\tilde{Y}) = \mathsf{E}[(X - \mu_X)(Y - \mu_Y)].$$

The Schwarz inequality further justifies expanding out the product in the final [ ] bracket to yield the alternative formula:

(c) $$\mathrm{Cov}(X, Y) = \mathsf{E}(XY) - \mu_X \mu_Y.$$

As you know, the *variance* of X is defined by

(d) $$\mathrm{Var}(X) := \mathsf{E}[(X - \mu_X)^2] = \mathsf{E}(X^2) - \mu_X^2 = \mathrm{Cov}(X, X).$$

Inner product, angle

For $U, V \in \mathcal{L}^2$, we define the *inner* (or *scalar*) *product*

(e) $$\langle U, V \rangle := \mathsf{E}(UV),$$

and if $\|U\|_2$ and $\|V\|_2 \neq 0$, we define the cosine of the angle θ between U and V by

(f) $$\cos \theta = \frac{\langle U, V \rangle}{\|U\|_2 \|V\|_2}.$$

This has modulus at most 1 by the Schwarz inequality. This ties in with the probabilistic idea of correlation:

(g) *the correlation ρ of X and Y is $\cos \alpha$ where α is the angle between $\tilde{X}$ and $\tilde{Y}$.*

Orthogonality, Pythagoras theorem

$\mathcal{L}^2$ has the same geometry as any inner-product space (but see 'Quotienting' below). Thus the 'cosine rule' of elementary geometry holds, and the Pythagoras theorem takes the form

(h) $$\|U + V\|_2{}^2 = \|U\|_2{}^2 + \|V\|_2{}^2 \text{ if } \langle U, V \rangle = 0.$$

If $\langle U, V \rangle = 0$, we say that U and V are *orthogonal* or *perpendicular*, and write $U \perp V$. In probabilistic language, (h) takes the form (with U, V replaced by $\tilde{X}, \tilde{Y}$)

(i) $$\mathrm{Var}(X + Y) = \mathrm{Var}(X) + \mathrm{Var}(Y) \text{ if } \mathrm{Cov}(X, Y) = 0.$$

Generally, for $X_1, X_2, \ldots, X_n \in \mathcal{L}^2$,

(j) $$\mathrm{Var}(X_1 + X_2 + \cdots + X_n) = \sum_k \mathrm{Var}(X_k) + 2 \sum \sum_{i<j} \mathrm{Cov}(X_i, X_j).$$

I have not marked results such as (i) and (j) with ▶ because I am sure that they are well known to you.

Parallelogram law

Note that by the bilinearity of $\langle \cdot, \cdot \rangle$,

(k) $$\|U + V\|_2{}^2 + \|U - V\|_2{}^2 = \langle U + V, U + V \rangle + \langle U - V, U - V \rangle$$
$$= 2\|U\|_2{}^2 + 2\|V\|_2{}^2.$$

Quotienting (or lack of it!): L^2

Our space $\mathcal{L}^2$ does not quite satisfy the requirements for an inner product space because the best we can say is that (see (5.2,b))

$$\|U\|_2 = 0 \text{ if and only if } U = 0 \text{ almost surely.}$$

In functional analysis, we find an elegant solution by defining an equivalence relation

$$U \sim V \text{ if and only if } U = V \text{ almost surely}$$

and define L^2 as '$\mathcal{L}^2$ quotiented out by this equivalence relation'. Of course, one needs to check that if for $i = 1, 2$, we have $c_i \in \mathbb{R}$ and $U_i, V_i \in \mathcal{L}^2$ with $U_i \sim V_i$, then

$$c_1 U_1 + c_2 U_2 \sim c_1 V_1 + c_2 V_2; \qquad \langle U_1, U_2 \rangle = \langle V_1, V_2 \rangle;$$

that if $U_n \to U$ in $\mathcal{L}^2$ and $V_n \sim U_n$ and $V \sim U$, then $V_n \to V$ in $\mathcal{L}^2$; etc. Then L^2 is a true vector space, every equivalence class having a representative with all its values finite.

As mentioned in 'A Question of Terminology', we normally do not do this quotienting in probability theory. Although one might safely do so at the moderately elementary level of this book, one could not do so at a more advanced level. For a Brownian motion $\{B_t : t \in \mathbb{R}^+\}$, the crucial property that $t \mapsto B_t(\omega)$ is continuous would be meaningless if one replaced the true function B_t on Ω by an equivalence class.

6.10. Completeness of $\mathcal{L}^p$ $(1 \leq p < \infty)$

Let $p \in [1, \infty)$.

The following result (a) is important in functional analysis, and will be crucial for us in the case when $p = 2$. It is instructive to prove it as an exercise in our probabilistic way of thinking, and we now do so.

(a)   *If (X_n) is a Cauchy sequence in $\mathcal{L}^p$ in that*

$$\sup_{r, s \geq k} \|X_r - X_s\|_p \to 0 \qquad (k \to \infty)$$

then there exists X in $\mathcal{L}^p$ such that $X_r \to X$ in $\mathcal{L}^p$:

$$\|X_r - X\|_p \to 0 \qquad (r \to \infty).$$

Note. Property (a) is important in showing that $\mathcal{L}^p$ can be made into a *Banach space* L^p by a quotienting technique of the type mentioned at the end of the preceding section.

Proof of (a). We show that X may be chosen to be an almost sure limit of a subsequence (X_{k_n}).

Choose a sequence $(k_n : n \in \mathsf{N})$ with $k_n \uparrow \infty$ such that

$$(r, s \geq k_n) \quad \Rightarrow \quad \|X_r - X_s\|_p < 2^{-n}.$$

Then

$$\mathsf{E}(|X_{k_{n+1}} - X_{k_n}|) = \|X_{k_{n+1}} - X_{k_n}\|_1 \leq \|X_{k_{n+1}} - X_{k_n}\|_p < 2^{-n},$$

so that

$$\mathsf{E} \sum |X_{k_{n+1}} - X_{k_n}| < \infty.$$

Hence it is almost surely true that the series

$$\sum (X_{k_{n+1}} - X_{k_n}) \quad \text{converges}$$

(even absolutely!), so that

$$\lim X_{k_n}(\omega) \text{ exists for almost all } \omega.$$

Define

$$X(\omega) := \limsup X_{k_n}(\omega), \forall \omega.$$

Then X is $\mathcal{F}$-measurable, and $X_{k_n} \to X$, a.s.

Suppose that $n \in \mathsf{N}$ and $r \geq k_n$. Then, for $\mathsf{N} \ni t \geq n$,

$$\mathsf{E}(|X_r - X_{k_t}|^p) = \|X_r - X_{k_t}\|_p^{\,p} \leq 2^{-np},$$

so that on letting $t \uparrow \infty$ and using Fatou's Lemma, we obtain

$$\mathsf{E}(|X_r - X|^p) \leq 2^{-np}.$$

Firstly, $X_r - X \in \mathcal{L}^p$, so that $X \in \mathcal{L}^p$. Secondly, we see that, indeed, $X_r \to X$ in $\mathcal{L}^p$. $\qquad\qquad\square$

Note. For an easy exercise on $\mathcal{L}^p$ convergence, see EA13.2.

6.11. Orthogonal projection

The result on completeness of $\mathcal{L}^p$ obtained in the previous section has a number of important consequences for probability theory, and it is perhaps as well to develop one of these while Section 6.10 is fresh in your mind.

I hope that you will allow me to present the following result on orthogonal projection as a piece of geometry for now, deferring discussion of its central rôle in the theory of *conditional expectation* until Chapter 9.

We write $\| \cdot \|$ for $\| \cdot \|_2$ throughout this section.

THEOREM

▶   *Let $\mathcal{K}$ be a vector subspace of $\mathcal{L}^2$ which is complete in that whenever (V_n) is a sequence in $\mathcal{K}$ which has the Cauchy property that*

$$\sup_{r,s \geq k} \|V_r - V_s\| \to 0 \qquad (k \to \infty),$$

then there exists a V in $\mathcal{K}$ such that

$$\|V_n - V\| \to 0 \qquad (n \to \infty).$$

Then given X in $\mathcal{L}^2$, there exists Y in $\mathcal{K}$ such that

(i) $\|X - Y\| = \Delta := \inf\{\|X - W\| : W \in \mathcal{K}\},$

(ii) $X - Y \perp Z, \quad \forall Z \in \mathcal{K}.$

Properties (i) and (ii) of Y in $\mathcal{K}$ are equivalent and if $\tilde{Y}$ shares either property (i) or (ii) with Y, then

$$\|\tilde{Y} - Y\| = 0 \quad (equivalently, \quad Y = \tilde{Y}, \text{a.s.}).$$

Definition. The random variable Y in the theorem is called a version of the *orthogonal projection of X onto $\mathcal{K}$*. If $\tilde{Y}$ is another version, then $\tilde{Y} = Y$, a.s.

Proof. Choose a sequence (Y_n) in $\mathcal{K}$ such that

$$\|X - Y_n\| \to \Delta.$$

By the parallelogram law (6.9,k),

$$\|X - Y_r\|^2 + \|X - Y_s\|^2 = 2\|X - \tfrac{1}{2}(Y_r + Y_s)\|^2 + 2\|\tfrac{1}{2}(Y_r - Y_s)\|^2.$$

But $\tfrac{1}{2}(Y_r + Y_s) \in \mathcal{K}$, so that $\|X - \tfrac{1}{2}(Y_r + Y_s)\|^2 \geq \Delta^2$. It is now obvious that the sequence (Y_n) has the Cauchy property so that there exists a Y in $\mathcal{K}$ such that

$$\|Y_n - Y\| \to 0.$$

Since (6.8,b) implies that $\|X - Y\| \leq \|X - Y_n\| + \|Y_n - Y\|$, it is clear that

$$\|X - Y\| = \Delta.$$

For any Z in $\mathcal{K}$, we have $Y + tZ \in \mathcal{K}$ for $t \in \mathbf{R}$, and so

$$\|X - Y - tZ\|^2 \geq \|X - Y\|^2,$$

whence

$$-2t\langle Z, X - Y \rangle + t^2 \|Z\|^2 \geq 0.$$

This can only be the case for all t of small modulus if

$$\langle Z, X - Y \rangle = 0. \qquad \square$$

Remark. The case to which we shall apply this theorem is when $\mathcal{K}$ has the form $\mathcal{L}^2(\Omega, \mathcal{G}, \mathbf{P})$ for some sub-σ-algebra $\mathcal{G}$ of $\mathcal{F}$.

6.12. The 'elementary formula' for expectation

Back to earth!

Let X be a random variable. To avoid confusion between different $\mathcal{L}$'s, let us here write Λ_X on $(\mathbf{R}, \mathcal{B})$ for the law of X:

$$\Lambda_X(B) := \mathbf{P}(X \in B).$$

LEMMA

▶   *Suppose that h is a Borel measurable function from $\mathbf{R}$ to $\mathbf{R}$. Then*

$$h(X) \in \mathcal{L}^1(\Omega, \mathcal{F}, \mathbf{P}) \text{ if and only if } h \in \mathcal{L}^1(\mathbf{R}, \mathcal{B}, \Lambda_X)$$

and then

(a) $$\mathbf{E}h(X) = \Lambda_X(h) = \int_{\mathbf{R}} h(x)\Lambda_X(dx).$$

Proof. We simply feed everything into the standard machine.

Result (a) is the *definition* of Λ_X if $h = I_B$ ($B \in \mathcal{B}$). Linearity then shows that (a) is true if h is a simple function on $(\mathbf{R}, \mathcal{B})$. (MON) then implies (a) for h a non-negative function, and linearity allows us to complete the argument. $\qquad \square$

Probability density function (pdf)

We say that X has a probability density function (pdf) f_X if there exists a Borel function $f_X : \mathbf{R} \to [0, \infty]$ such that

(b) $$\mathbf{P}(X \in B) = \int_B f_X(x)dx, \qquad B \in \mathcal{B}.$$

Here we have written dx for what should be Leb(dx). In the language of Section 5.12, result (b) says that Λ_X has density f_X relative to Leb:

$$\frac{d\Lambda_X}{d\text{Leb}} = f_X.$$

The function f_X is only defined almost everywhere: any function a.e. equal to f_X will also satisfy (b) 'and conversely'.

The above lemma extends to

$$E(|h(X)|) < \infty \text{ if and only if } \int |h(x)| f_X(x) dx < \infty$$

and then

$$E h(X) = \int_{\mathbf{R}} h(x) f_X(x) dx.$$

6.13. Hölder from Jensen

The truncation technique used to prove the Schwarz inequality in Section 6.8 relied on the fact that $P(\Omega) < \infty$. However, the Schwarz inequality is true for any measure space, as is the more general Hölder inequality.

We conclude this chapter with a device (often useful) which yields the Hölder inequality for any (S, Σ, μ) from Jensen's inequality for probability triples.

Let (S, Σ, μ) be a measure space. Suppose that

▶
$$p > 1 \text{ and } p^{-1} + q^{-1} = 1.$$

Write $f \in \mathcal{L}^p(S, \Sigma, \mu)$ if $f \in m\Sigma$ and $\mu(|f|^p) < \infty$, and in that case define

$$\|f\|_p := \{\mu(|f|^p)\}^{1/p}.$$

THEOREM

Suppose that $f, g \in \mathcal{L}^p(S, \Sigma, \mu)$, $h \in \mathcal{L}^q(S, \Sigma, \mu)$. Then

▶(a)   (**Hölder's inequality**) $fh \in \mathcal{L}^1(S, \Sigma, \mu)$ *and*

$$|\mu(fh)| \leq \mu(|fh|) \leq \|f\|_p \|h\|_q;$$

▶(b)   (**Minkowski's inequality**)

$$\|f + g\|_p \leq \|f\|_p + \|g\|_p.$$

Proof of (a). We can obviously restrict attention to the case when

$$f, h \geq 0 \text{ and } \mu(f^p) > 0.$$

With the notation of Section 5.14, define

$$\mathbf{P} := \frac{f^p \mu}{\mu(f^p)},$$

so that $\mathbf{P}$ is a probability measure on (S, Σ). Define

$$u(s) := \begin{cases} h(s)/f(s)^{p-1} & \text{if } f(s) > 0, \\ 0 & \text{if } f(s) = 0. \end{cases}$$

The fact that $\mathbf{P}(u)^q \leq \mathbf{P}(u^q)$ now yields

$$\mu(|fh|) \leq \|f\|_p \|h I_{\{f>0\}}\|_q \leq \|f\|_p \|h\|_q. \qquad \square$$

Proof of (b). Using Hölder's inequality, we have

$$\mu(|f+g|^p) = \mu(|f||f+g|^{p-1}) + \mu(|g||f+g|^{p-1})$$
$$\leq \|f\|_p A + \|g\|_p A,$$

where

$$A = \||f+g|^{p-1}\|_q = \mu(|f+g|^p)^{1/q},$$

and (b) follows on rearranging. (The result is non-trivial only if $f, g \in \mathcal{L}^p$, and in that case, the finiteness of A follows from the vector-space property of $\mathcal{L}^p$.) $\qquad \square$

Chapter 7
An Easy Strong Law

7.1. 'Independence means multiply' – again!

THEOREM

▶ *Suppose that X and Y are independent RVs, and that X and Y are both in $\mathcal{L}^1$. Then $XY \in \mathcal{L}^1$ and*

$$E(XY) = E(X)E(Y).$$

In particular, if X and Y are independent elements of $\mathcal{L}^2$, then

$$\mathrm{Cov}(X,Y) = 0 \text{ and } \mathrm{Var}(X+Y) = \mathrm{Var}(X) + \mathrm{Var}(Y).$$

Proof. Writing $X = X^+ - X^-$, etc., allows us to reduce the problem to the case when $X \geq 0$ and $Y \geq 0$. This we do.

But then, if $\alpha^{(r)}$ is our familiar staircase function, then

$$\alpha^{(r)}(X) = \sum a_i I_{A_i}, \qquad \alpha^{(r)}(Y) = \sum b_j I_{B_j}$$

where the sums are over finite parameter sets, and where for each i and j, A_i (in $\sigma(X)$) is independent of B_j (in $\sigma(Y)$). Hence

$$
\begin{aligned}
E[\alpha^{(r)}(X)\alpha^{(r)}(Y)] &= \sum\sum a_i b_j P(A_i \cap B_j) \\
&= \sum\sum a_i b_j P(A_i)P(B_j) = E[\alpha^{(r)}(X)]E[\alpha^{(r)}(Y)].
\end{aligned}
$$

Now let $r \uparrow \infty$ and use (MON). □

Remark. Note especially that if X and Y are independent then $X \in \mathcal{L}^1$ and $Y \in \mathcal{L}^1$ imply that $XY \in \mathcal{L}^1$. This is not necessarily true when X and

Y are not independent, and we need the inequalities of Schwarz, Hölder, etc. It is important that independence obviates the need for such inequalities.

7.2. Strong Law – first version

The following result covers many cases of importance. You should note that though it imposes a 'finite 4th moment' condition, it makes no assumption about identical distributions for the (X_n) sequence. It is remarkable that so fine a result has so simple a proof.

THEOREM

▶   *Suppose that $X_1, X_2, \ldots$ are independent random variables, and that for some constant K in $[0, \infty)$,*

$$\mathsf{E}(X_k) = 0, \quad \mathsf{E}(X_k^4) \leq K, \quad \forall k.$$

Let $S_n = X_1 + X_2 + \cdots + X_n$. Then

$$\mathsf{P}(n^{-1} S_n \to 0) = 1,$$

or again, $S_n/n \to 0$ (a.s.).

Proof. We have

$$\mathsf{E}(S_n^4) = \mathsf{E}[(X_1 + X_2 + \cdots + X_n)^4]$$
$$= \mathsf{E}(\sum_k X_k^4 + 6 \sum \sum_{i<j} X_i^2 X_j^2),$$

because, for distinct i, j, k and l,

$$\mathsf{E}(X_i X_j^3) = \mathsf{E}(X_i X_j^2 X_k) = \mathsf{E}(X_i X_j X_k X_l) = 0,$$

using independence plus the fact that $\mathsf{E}(X_i) = 0$. [Note that, for example, the fact that $\mathsf{E}(X_j^4) < \infty$ implies that $\mathsf{E}(X_j^3) < \infty$, by the 'monotonicity of $\mathcal{L}^p$ norms' result in Section 6.7. Thus X_i and X_j^3 are in $\mathcal{L}^1$.]

We know from Section 6.7 that

$$[\mathsf{E}(X_i^2)]^2 \leq \mathsf{E}(X_i^4) \leq K, \quad \forall i.$$

Hence, using independence again, for $i \neq j$,

$$\mathsf{E}(X_i^2 X_j^2) = \mathsf{E}(X_i^2)\mathsf{E}(X_j^2) \leq K.$$

Thus

$$E(S_n^4) \le nK + 3n(n-1)K \le 3Kn^2,$$

and (see Section 6.5)

$$E \sum (S_n/n)^4 \le 3K \sum n^{-2} < \infty,$$

so that $\sum (S_n/n)^4 < \infty$, a.s., and

$$S_n/n \to 0, \quad \text{a.s.} \qquad \square$$

Corollary. *If the condition* $E(X_k) = 0$ *in the theorem is replaced by* $E(X_k) = \mu$ *for some constant* μ*, then the theorem holds with* $n^{-1}S_n \to \mu$ *(a.s.) as its conclusion.*

Proof. It is obviously a case of applying the theorem to the sequence (Y_k), where $Y_k := X_k - \mu$. But we need to know that

(a) $$\sup_k E(Y_k^4) < \infty.$$

This is obvious from Minkowski's inequality

$$\|X_k - \mu\|_4 \le \|X_k\|_4 + |\mu|$$

(the constant function $\mu 1$ on Ω having $\mathcal{L}^4$ norm $|\mu|$). But we can also prove (a) immediately by the elementary inequality (6.7,b). $\qquad \square$

___

The next topics indicate a different use of variance.

___

7.3. Chebyshev's inequality

As you know this says that *for* $c \ge 0$*, and* $X \in \mathcal{L}^2$,

$$c^2 P(|X - \mu| > c) \le \text{Var}(X), \qquad \mu := E(X);$$

and it is obvious.

Example. Consider a sequence (X_n) of IID RVs with values in $\{0,1\}$ with

$$p = P(X_n = 1) = 1 - P(X_n = 0).$$

Then $E(X_n) = p$ and $\text{Var}(X_n) = p(1-p) \leq \frac{1}{4}$. Thus (using Theorem 7.1)

$$S_n := X_1 + X_2 + \cdots + X_n$$

has expectation np and variance $np(1-p) \leq n/4$, and we have

$$E(n^{-1}S_n) = p, \quad \text{Var}(n^{-1}S_n) = n^{-2}\text{Var}(S_n) \leq 1/(4n).$$

Chebyshev's inequality yields

$$P(|n^{-1}S_n - p| > \delta) \leq 1/(4n\delta^2).$$

7.4. Weierstrass approximation theorem

If f is a continuous function on $[0,1]$ and $\varepsilon > 0$, then there exists a polynomial B such that

$$\sup_{x \in [0,1]} |B(x) - f(x)| \leq \varepsilon.$$

Proof. Let (X_k), S_n etc. be as in the Example in Section 7.3. You are well aware that

$$P[S_n = k] = \binom{n}{k} p^k (1-p)^{n-k}, \quad 0 \leq k \leq n.$$

Hence

$$B_n(p) := Ef(n^{-1}S_n) = \sum_{k=0}^{n} f(n^{-1}k)\binom{n}{k}p^k(1-p)^{n-k},$$

the 'B' being in deference to Bernstein.

Now f is *bounded* on $[0,1]$, $|f(y)| \leq K$, $\forall y \in [0,1]$. Also, f is *uniformly continuous* on $[0,1]$: for our given $\varepsilon > 0$, there exists $\delta > 0$ such that

(a) $|x - y| \leq \delta$ implies that $|f(x) - f(y)| < \frac{1}{2}\varepsilon$.

Now, for $p \in [0,1]$,

$$|B_n(p) - f(p)| = |E\{f(n^{-1}S_n) - f(p)\}|.$$

Let us write $Y_n := |f(n^{-1}S_n) - f(p)|$ and $Z_n := |n^{-1}S_n - p|$. Then $Z_n \leq \delta$ implies that $Y_n < \frac{1}{2}\varepsilon$, and we have

$$|B_n(p) - f(p)| \leq E(Y_n)$$
$$= E(Y_n; Z_n \leq \delta) + E(Y_n; Z_n > \delta)$$
$$\leq \tfrac{1}{2}\varepsilon P(Z_n \leq \delta) + 2KP(Z_n > \delta)$$
$$\leq \tfrac{1}{2}\varepsilon + 2K/(4n\delta^2).$$

Earlier, we chose a fixed δ at (a). We now choose n so that

$$2K/(4n\delta^2) < \tfrac{1}{2}\varepsilon.$$

Then $|B_n(p) - f(p)| < \varepsilon$, for all p in $[0,1]$.   □

Now do Exercise E7.1 on inverting Laplace transforms.

Chapter 8
Product Measure

8.0. Introduction and advice

One of this chapter's main lessons of practical importance is that **an 'interchange of order of integration' result**

$$\int_{S_1}\left(\int_{S_2} f(s_1,s_2)\,\mu_2(ds_2)\right)\mu_1(ds_1) = \int_{S_2}\left(\int_{S_1} f(s_1,s_2)\,\mu_1(ds_1)\right)\mu_2(ds_2)$$

is always valid (both sides possibly being infinite) if $f \geq 0$; and is valid for 'signed' f (with both repeated integrals finite) provided that one (then the other) of the integrals of absolute values:

$$\int_{S_1}\left(\int_{S_2} |f(s_1,s_2)|\,\mu_2(ds_2)\right)\mu_1(ds_1) = \int_{S_2}\left(\int_{S_1} |f(s_1,s_2)|\,\mu_1(ds_1)\right)\mu_2(ds_2)$$

is finite.

It is a good idea to read through the chapter to get the ideas, but you are strongly recommended to postpone serious study of the contents until a later stage. Except for the matter of infinite products, it is all a case of relentless use of either the standard machine or the Monotone-Class Theorem to prove intuitively obvious things made to look complicated by the notation. When you do begin a serious study, it is important to appreciate when the more subtle Monotone-Class Theorem has to be used instead of the standard machine.

8.1. Product measurable structure, $\Sigma_1 \times \Sigma_2$

Let (S_1, Σ_1) and (S_2, Σ_2) be measurable spaces. Let S denote the Cartesian product $S := S_1 \times S_2$. For $i = 1, 2$, let ρ_i denote the ith coordinate map, so that

$$\rho_1(s_1, s_2) := s_1, \qquad \rho_2(s_1, s_2) := s_2.$$

The fundamental definition of $\Sigma = \Sigma_1 \times \Sigma_2$ is as the σ-algebra

▶(a) $\Sigma = \sigma(\rho_1, \rho_2).$

Thus Σ is generated by the sets of the form

$$\rho_1^{-1}(B_1) = B_1 \times S_2 \qquad (B_1 \in \Sigma_1)$$

together with sets of the form

$$\rho_2^{-1}(B_2) = S_1 \times B_2 \qquad (B_2 \in \Sigma_2).$$

Generally, *a product σ-algebra is generated by Cartesian products in which one factor is allowed to vary over the σ-algebra corresponding to that factor, and all other factors are whole spaces.* In the case of our product of *two* factors, we have

(b) $$(B_1 \times S_2) \cap (S_1 \times B_2) = B_1 \times B_2$$

and you can easily check that

(c) $$\mathcal{I} = \{B_1 \times B_2 : B_i \in \Sigma_i\}$$

is a π-system generating $\Sigma = \Sigma_1 \times \Sigma_2$. A similar remark would apply for a *countable* product $\prod \Sigma_n$, but you can see that, since we may only take *countable* intersections in analogues of (b), products of uncountable families of σ-algebras cause problems. The fundamental definition analogous to (a) still works.

LEMMA

(d)   *Let $\mathcal{H}$ denote the class of functions $f : S \to \mathbf{R}$ which are in $b\Sigma$ and which are such that*

for each s_1 in S_1, the map $s_2 \mapsto f(s_1, s_2)$ is Σ_2-measurable on S_2,
for each s_2 in S_2, the map $s_1 \mapsto f(s_1, s_2)$ is Σ_1-measurable on S_1.

Then $\mathcal{H} = b\Sigma$.

Proof. It is clear that if $A \in \mathcal{I}$, then $I_A \in \mathcal{H}$. Verification that $\mathcal{H}$ satisfies the conditions (i)-(iii) of the Monotone-Class Theorem 3.14 is straightforward. Since $\Sigma = \sigma(\mathcal{I})$, the result follows.   □

8.2. Product measure, Fubini's Theorem

We continue with the notation of the preceding Section. We suppose that for $i = 1, 2$, μ_i is a *finite* measure on (S_i, Σ_i). We know from the preceding Section that for $f \in b\Sigma$, we may define the integrals

$$I_1^f(s_1) := \int_{S_2} f(s_1, s_2) \mu_2(ds_2), \quad I_2^f(s_2) := \int_{S_1} f(s_1, s_2) \mu_1(ds_1).$$

LEMMA

> *Let $\mathcal{H}$ be the class of elements in* $b\Sigma$ *such that the following property holds:*
>
> $$I_1^f(\cdot) \in b\Sigma_1 \text{ and } I_2^f(\cdot) \in b\Sigma_2 \text{ and}$$
>
> $$\int_{S_1} I_1^f(s_1)\mu_1(ds_1) = \int_{S_2} I_2^f(s_2)\mu_2(ds_2).$$
>
> *Then $\mathcal{H} = b\Sigma$.*

Proof. If $A \in \mathcal{I}$, then, trivially, $I_A \in \mathcal{H}$. Verification of the conditions of Monotone-Class Theorem 3.14 is straightforward.   □

For $F \in \Sigma$ with indicator function $f := I_F$, we now define

$$\mu(F) := \int_{S_1} I_1^f(s_1)\mu_1(ds_1) = \int_{S_2} I_2^f(s_2)\mu_2(ds_2).$$

Fubini's Theorem

▶▶   *The set function μ is a measure on (S, Σ) called the* **product measure** *of μ_1 and μ_2 and we write $\mu = \mu_1 \times \mu_2$ and*

$$(S, \Sigma, \mu) = (S_1, \Sigma_1, \mu_1) \times (S_2, \Sigma_2, \mu_2).$$

Moreover, μ is the unique measure on (S, Σ) for which

(a) $$\mu(A_1 \times A_2) = \mu_1(A_1)\mu_2(A_2), \qquad A_i \in \Sigma_i.$$

If $f \in (m\Sigma)^+$, then with the obvious definitions of I_1^f, I_2^f, we have

(b) $$\mu(f) = \int_{S_1} I_1^f(s_1)\mu_1(ds_1) = \int_{S_2} I_2^f(s_2)\mu_2(ds_2),$$

in $[0, \infty]$. If $f \in \mathrm{m}\Sigma$ and $\mu(|f|) < \infty$, then equation (a) is valid (with all terms in $\mathbf{R}$).

Proof. The fact that μ is a measure is a consequence of linearity and (MON). The fact that μ is then uniquely specified by (a) is obvious from Uniqueness Lemma 1.6 and the fact that $\sigma(\mathcal{I}) = \Sigma$.

Result (b) is automatic for $f = I_A$, where $A \in \mathcal{I}$. The Monotone-Class Theorem shows that it is therefore valid for $f \in \mathrm{b}\Sigma$, and in particular for f in the SF^+ space for (S, Σ, μ). (MON) then shows that it is valid for $f \in (\mathrm{m}\Sigma)^+$; and linearity shows that (b) is valid if $\mu(|f|) < \infty$.

Extension

▸   *All of Fubini's Theorem will work if the (S_i, Σ_i, μ_i) are σ-finite measure spaces:*

We have a unique measure μ on (S, Σ) satisfying (a), etc., etc. We can prove this by breaking up σ-finite spaces into countable unions of disjoint finite blocks.

Warning

The σ-finiteness condition cannot be dropped. The standard example is the following. For $i = 1, 2$, take $S_i = [0, 1]$ and $\Sigma_i = \mathcal{B}[0, 1]$. Let μ_1 be Lebesgue measure and let μ_2 just count the number of elements in a set. Let F be the diagonal $\{(x, y) \in S_1 \times S_2 : x = y\}$. Then (check!) $F \in \Sigma$, but

$$I_1^f(s_1) \equiv 1, \quad I_2^f(s_2) \equiv 0$$

and result (b) fails, stating that $1 = 0$.

Something to think about

So, our insistence on beginning with *bounded* functions on products of *finite* measures was necessary. Perhaps it is worth emphasizing that in our standard machine, things work because we can use indicator functions of any set in our σ-algebra, whereas when we can only use indicator functions of sets in a π-system, we have to use the Monotone-Class Theorem. We cannot approximate the set F in the Warning example as

$$F = \uparrow \lim F_n,$$

where each F_n is a finite union of 'rectangles' $A_1 \times A_2$, each A_i being in $\mathcal{B}[0, 1]$.

A simple application

Suppose that X is a non-negative random variable on $(\Omega, \mathcal{F}, \mathbf{P})$. Consider the measure $\mu := \mathbf{P} \times \text{Leb}$ on $(\Omega, \mathcal{F}) \times ([0, \infty), \mathcal{B}[0, \infty))$. Let

$$A := \{(\omega, x) : 0 \le x \le X(\omega)\}, \quad h := I_A.$$

Note that A is the 'region under the graph of X'. Then

$$I_1^h(\omega) = X(\omega), \quad I_2^h(x) = \mathbf{P}(X \ge x).$$

Thus

(c)
$$\mu(A) = \mathbf{E}(X) = \int_{[0,\infty)} \mathbf{P}(X \ge x) dx,$$

dx denoting $\text{Leb}(dx)$ as usual. Thus we have obtained one of the well-known formulae for $\mathbf{E}(X)$ and also interpreted the integral $\mathbf{E}(X)$ as '*area under the graph of X*'.

Note. It is perhaps worth remarking that the Monotone-Class Theorem, the Fatou Lemma and the reverse Fatou Lemma for functions amount to the corresponding results for sets applied to regions under graphs.

8.3. Joint laws, joint pdfs

Let X and Y be two random variables. The *(joint) law* $\mathcal{L}_{X,Y}$ of the pair (X, Y) is the map

$$\mathcal{L}_{X,Y} : \mathcal{B}(\mathbf{R}) \times \mathcal{B}(\mathbf{R}) \to [0, 1]$$

defined by

$$\mathcal{L}_{X,Y}(\Gamma) := \mathbf{P}[(X, Y) \in \Gamma].$$

The system $\{(-\infty, x] \times (-\infty, y] : x, y \in \mathbf{R}\}$ is a π-system which generates $\mathcal{B}(\mathbf{R}) \times \mathcal{B}(\mathbf{R})$. Hence $\mathcal{L}_{X,Y}$ is completely determined by the joint distribution $F_{X,Y}$ of X and Y which is defined via

$$F_{X,Y}(x, y) := \mathbf{P}(X \le x; Y \le y).$$

We now know how to construct Lebesgue measure

$$\mu = \text{Leb} \times \text{Leb on } (\mathbf{R}, \mathcal{B}(\mathbf{R}))^2.$$

We say that X and Y have joint probability density function (joint pdf) $f_{X,Y}$ on $\mathbf{R}^2$ if for $\Gamma \in \mathcal{B}(\mathbf{R}) \times \mathcal{B}(\mathbf{R})$,

$$\mathbf{P}[(X, Y) \in \Gamma] = \int_\Gamma f_{X,Y}(z)\mu(dz)$$

$$= \int_{\mathbf{R}} \int_{\mathbf{R}} I_\Gamma(x, y) f_{X,Y}(x, y) dx dy.$$

etc., etc., (Fubini's Theorem being used in the last step(s)). Fubini's Theorem further shows that

$$f_X(x) := \int_{\mathbb{R}} f_{X,Y}(x,y)dy$$

acts as a pdf for X (Section 6.12), etc., etc. You don't need me to tell you any more of this sort of thing.

8.4. Independence and product measure

Let X and Y be two random variables with laws $\mathcal{L}_X, \mathcal{L}_Y$ respectively and distribution functions F_X, F_Y respectively. Then the following three statements are equivalent:

  (i) *X and Y are independent;*

  (ii) $\mathcal{L}_{X,Y} = \mathcal{L}_X \times \mathcal{L}_Y$;

  (iii) $F_{X,Y}(x,y) = F_X(x)F_Y(y)$;

moreover, if (X,Y) has 'joint' pdf $f_{X,Y}$ then each of (i)-(iii) is equivalent to

  (iv) $f_{X,Y}(x,y) = f_X(x)f_Y(y)$ *for* Leb $\times$ Leb *almost every* (x,y).

You do not wish to know more about this either.

8.5. $\mathcal{B}(\mathbb{R})^n = \mathcal{B}(\mathbb{R}^n)$

Here again, things are nice and tidy provided we work with finite or countable products, but require different concepts (such as Baire σ-algebras) if we work with uncountable products.

$\mathcal{B}(\mathbb{R}^n)$ is constructed from the topological space $\mathbb{R}^n$. Now, if $\rho_i : \mathbb{R}^n \to \mathbb{R}$ is the ith coordinate map:

$$\rho_i(z_1, z_2, \ldots, z_n) = z_i,$$

then ρ_i is continuous, and hence $\mathcal{B}(\mathbb{R}^n)$-measurable. Hence

$$\mathcal{B}^n :=: \mathcal{B}(\mathbb{R})^n = \sigma(\rho_i : 1 \leq i \leq n) \subseteq \mathcal{B}(\mathbb{R}^n).$$

On the other hand, $\mathcal{B}(\mathbb{R}^n)$ is generated by the open subsets of $\mathbb{R}^n$, and every such open subset is a countable union of open 'hypercubes' of the form

$$\prod_{1 \leq i \leq n} (a_i, b_i)$$

and such products are in $\mathcal{B}(\mathbb{R})^n$. Hence, $\mathcal{B}(\mathbb{R}^n) = \mathcal{B}(\mathbb{R})^n$.   □

    In probability theory it is almost always product structures $\mathcal{B}^n$ which feature rather than $\mathcal{B}(\mathbb{R}^n)$. See Section 8.8.

8.6. The n-fold extension

So far in this chapter, we have studied the product measure space of *two* measure spaces and how this relates to the study of *two* random variables. You are more than able to 'generalize' 'from two to n' from your experience of similar things in other branches of mathematics. You should give some thought to the associativity of the 'product' in product measure space, something again familiar in analogous contexts.

8.7. Infinite products of probability triples

This topic is *not* a trivial extension of previous results. We concentrate on a restricted context (though an important one) because it allows us to get the main idea in a clear fashion; and extension to infinite products of arbitrary probability triples is then a purely routine exercise.

Canonical model for a sequence of independent RVs

Let $(\Lambda_n : n \in \mathbf{N})$ be a sequence of probability measures on $(\mathbf{R}, \mathcal{B})$. We already know from the coin-tossing trickery of Section 4.6 that we can construct a sequence (X_n) of independent RVs, X_n having law Λ_n. Here is a more elegant and systematic way of doing this.

THEOREM

> Let $(\Lambda_n : n \in \mathbf{N})$ be a sequence of probability measures on $(\mathbf{R}, \mathcal{B})$. Define
>
> $$\Omega = \prod_{n \in \mathbf{N}} \mathbf{R}$$
>
> so that a typical element ω of Ω is a sequence (ω_n) in $\mathbf{R}$. Define
>
> $$X_n : \Omega \to \mathbf{R}, \qquad X_n(\omega) := \omega_n,$$
>
> and let $\mathcal{F} := \sigma(X_n : n \in \mathbf{N})$. Then there exists a unique probability measure $\mathbf{P}$ on $(\Omega, \mathcal{F})$ such that for $r \in \mathbf{N}$ and $B_1, B_2, \ldots, B_r \in \mathcal{B}$,
>
> (a)
> $$\mathbf{P}\left(\left(\prod_{1 \leq k \leq r} B_k \right) \times \prod_{k > r} \mathbf{R} \right) = \prod_{1 \leq k \leq r} \Lambda_k(B_k).$$
>
> We write
>
> $$(\Omega, \mathcal{F}, \mathbf{P}) = \prod_{n \in \mathbf{N}} (\mathbf{R}, \mathcal{B}, \Lambda_n).$$
>
> Then the sequence $(X_n : n \in \mathbf{N})$ is a sequence of independent RVs on $(\Omega, \mathcal{F}, \mathbf{P})$, X_n having law Λ_n.

Remarks. (i) The uniqueness of **P** follows in the usual way from Lemma 1.6, because product sets of the form which appear on the left-hand side of (a) form a π-system generating $\mathcal{F}$.

(ii) We could rewrite (a) more neatly as

$$\mathbf{P}(\prod_{k\in\mathbf{N}} B_k) = \prod_{k\in\mathbf{N}} \Lambda_k(B_k).$$

To see this, use the monotone-convergence property (1.10,b) of measures.

Proof of the theorem is deferred to the Appendix to Chapter 9.

8.8. Technical note on the existence of joint laws

Let $(\Omega, \mathcal{F})$, (S_1, Σ_1) and (S_2, Σ_2) be measurable spaces. For $i = 1, 2$, let $X_i : \Omega \to S_i$ be such that $X_i^{-1} : \Sigma_i \to \mathcal{F}$. Define

$$S := S_1 \times S_2, \quad \Sigma := \Sigma_1 \times \Sigma_2, \quad X(\omega) := (X_1(\omega), X_2(\omega)) \in S.$$

Then (**Exercise**) $X^{-1} : \Sigma \to \mathcal{F}$, so that X is an (S, Σ)-valued random variable, and if **P** is a probability measure on Ω, we can talk about the law μ of X (equals the joint law of X_1 and X_2) on $(S, \Sigma) : \mu = \mathbf{P} \circ X^{-1}$ on Σ.

Suppose now that S_1 and S_2 are metrizable spaces and that $\Sigma_i = \mathcal{B}(S_i)$ $(i = 1, 2)$. Then S is a metrizable space under the product topology. If S_1 and S_2 are separable, then $\Sigma = \mathcal{B}(S)$, and there is no 'conflict'. However, if S_1 and S_2 are not separable, then $\mathcal{B}(S)$ may be strictly larger than Σ, X need not be an $(S, \mathcal{B}(S))$-valued random variable, and the joint law of X_1 and X_2 need not exist on $(S, \mathcal{B}(S))$.

It is perhaps as well to be warned of such things. Note that the separability of **R** *was* used in proving that $\mathcal{B}(\mathbf{R}^n) \subseteq \mathcal{B}^n$ in Section 8.5.

PART B: MARTINGALE THEORY

Chapter 9
Conditional Expectation

9.1. A motivating example

Suppose that $(\Omega, \mathcal{F}, \mathbf{P})$ is a probability triple and that X and Z are random variables,

> X taking the distinct values $x_1, x_2, \ldots, x_m$,
> Z taking the distinct values $z_1, z_2, \ldots, z_n$.

Elementary conditional probability:

$$\mathbf{P}(X = x_i | Z = z_j) := \mathbf{P}(X = x_i; Z = z_j)/\mathbf{P}(Z = z_j)$$

and elementary conditional expectation:

$$\mathbf{E}(X | Z = z_j) = \sum x_i \mathbf{P}(X = x_i | Z = z_j)$$

are familiar to you. The random variable $Y = \mathbf{E}(X|Z)$, the conditional expectation of X given Z, is defined as follows:

(a)   if $Z(\omega) = z_j$, then $Y(\omega) := \mathbf{E}(X|Z = z_j) =: y_j$ (say).

    It proves to be very advantageous to look at this idea in a new way. 'Reporting to us the value of $Z(\omega)$' amounts to partitioning Ω into 'Z-atoms' on which Z is constant:

$$\Omega \quad \boxed{Z = z_1} \boxed{Z = z_2} \boxed{\cdots} \boxed{Z = z_n}$$

The σ-algebra $\mathcal{G} = \sigma(Z)$ generated by Z consists of sets $\{Z \in B\}$, $B \in \mathcal{B}$, and therefore consists precisely of the 2^n possible unions of the n Z-atoms. It is clear from (a) that Y is constant on Z-atoms, or, to put it better,

(b) Y is $\mathcal{G}$-measurable.

Next, since Y takes the constant value y_j on the Z-atom $\{Z = z_j\}$, we have:

$$\int_{\{Z=z_j\}} Y\, d\mathbb{P} = y_j \mathbb{P}(Z = z_j) = \sum_i x_i \mathbb{P}(X = x_i | Z = z_j) \mathbb{P}(Z = z_j)$$

$$= \sum_i x_i \mathbb{P}(X = x_i; Z = z_j) = \int_{\{Z=z_j\}} X\, d\mathbb{P}.$$

If we write $G_j = \{Z = z_j\}$, this says $\mathbb{E}(Y I_{G_j}) = \mathbb{E}(X I_{G_j})$. Since for every G in $\mathcal{G}$, I_G is a sum of I_{G_j}'s, we have $\mathbb{E}(Y I_G) = \mathbb{E}(X I_G)$, or

(c) $$\int_G Y\, d\mathbb{P} = \int_G X\, d\mathbb{P}, \qquad \forall G \in \mathcal{G}.$$

Results (b) and (c) suggest the central definition of modern probability.

9.2. Fundamental Theorem and Definition (Kolmogorov, 1933)

▶▶▶   *Let $(\Omega, \mathcal{F}, \mathbb{P})$ be a triple, and X a random variable with $\mathbb{E}(|X|) < \infty$. Let $\mathcal{G}$ be a sub-σ-algebra of $\mathcal{F}$. Then there exists a random variable Y such that*

(a)   *Y is $\mathcal{G}$ measurable,*

(b) $\mathbb{E}(|Y|) < \infty$,

(c)   *for every set G in $\mathcal{G}$ (**equivalently, for every set G in some π-system which contains Ω and generates $\mathcal{G}$**), we have*

$$\int_G Y\, d\mathbb{P} = \int_G X\, d\mathbb{P}, \qquad \forall G \in \mathcal{G}.$$

*Moreover, if $\tilde{Y}$ is another RV with these properties then $\tilde{Y} = Y$, a.s., that is, $\mathbb{P}[\tilde{Y} = Y] = 1$. A random variable Y with properties (a)-(c) is called **a version of the conditional expectation** $\mathbb{E}(X|\mathcal{G})$ of X given $\mathcal{G}$, and we write $Y = \mathbb{E}(X|\mathcal{G})$, a.s.*

Two versions agree a.s., and when one has become familiar with the concept, one identifies different versions and speaks of *the* conditional expectation $\mathbb{E}(X|\mathcal{G})$. But you should think about the 'a.s.' throughout this course.

The theorem is proved in Section 9.5, except for the π-system assertion which you will find at Exercise E9.1.

▶**Notation.** We often write $\mathbb{E}(X|Z)$ for $\mathbb{E}(X|\sigma(Z))$, $\mathbb{E}(X|Z_1, Z_2, \ldots)$ for $\mathbb{E}(X|\sigma(Z_1, Z_2, \ldots))$, etc. That this is consistent with the elementary usage is apparent from Section 9.6 below.

9.3. The intuitive meaning

An experiment has been performed. The only information available to you regarding which sample point ω has been chosen is the set of values $Z(\omega)$ for every $\mathcal{G}$-measurable random variable Z. Then $Y(\omega) = E(X|\mathcal{G})(\omega)$ is the expected value of $X(\omega)$ given this information. The 'a.s.' ambiguity in the definition is something one has to live with in general, but it is sometimes possible to choose a canonical version of $E(X|\mathcal{G})$.

Note that if $\mathcal{G}$ is the trivial σ-algebra $\{\emptyset, \Omega\}$ (which contains no information), then $E(X|\mathcal{G})(\omega) = E(X)$ for all ω.

9.4. Conditional expectation as least-squares-best predictor

▶▶   *If $E(X^2) < \infty$, then the conditional expectation $Y = E(X|\mathcal{G})$ is a version of the orthogonal projection (see Section 6.11) of X onto $\mathcal{L}^2(\Omega, \mathcal{G}, P)$. Hence, Y is the least-squares-best $\mathcal{G}$-measurable predictor of X: amongst all $\mathcal{G}$-measurable functions (i.e. amongst all predictors which can be computed from the available information), Y minimizes*

$$E[(Y - X)^2].$$

No surprise then that conditional expectation (and the martingale theory which develops it) is crucial in filtering and control – of space-ships, of industrial processes, or whatever.

9.5. Proof of Theorem 9.2

The standard way to prove Theorem 9.2 (see Section 14.14) is via the *Radon-Nikodým theorem* described in Section 5.14. However, Section 9.4 suggests a much simpler approach, and this is what we now develop. We can then prove the general Radon-Nikodým theorem by martingale theory. See Section 14.13.

First we prove the almost sure uniqueness of a version of $E(X|\mathcal{G})$. Then we prove the existence of $E(X|\mathcal{G})$ when $X \in \mathcal{L}^2$; and finally, we prove the existence in general.

Almost sure uniqueness of $E(X|\mathcal{G})$

Suppose that $X \in \mathcal{L}^1$ and that Y and $\tilde{Y}$ are versions of $E(X|\mathcal{G})$. Then $Y, \tilde{Y} \in \mathcal{L}^1(\Omega, \mathcal{G}, P)$, and

$$E(Y - \tilde{Y}; G) = 0, \qquad \forall G \in \mathcal{G}.$$

Suppose that Y and $\tilde{Y}$ are not almost surely equal. We may assume that the labelling is such that $\mathbf{P}(Y > \tilde{Y}) > 0$. Since

$$\{Y > \tilde{Y} + n^{-1}\} \uparrow \{Y > \tilde{Y}\},$$

we see that $\mathbf{P}(Y - \tilde{Y} > n^{-1}) > 0$ for some n. But the set $\{Y - \tilde{Y} > n^{-1}\}$ is in $\mathcal{G}$, because Y and $\tilde{Y}$ are $\mathcal{G}$-measurable; and

$$\mathbf{E}(Y - \tilde{Y}; Y - \tilde{Y} > n^{-1}) \geq n^{-1}\mathbf{P}(Y - \tilde{Y} > n^{-1}) > 0,$$

a contradiction. Hence $Y = \tilde{Y}$, a.s.   □

Existence of $\mathbf{E}(X|\mathcal{G})$ for $X \in \mathcal{L}^2$

Suppose that $X \in \mathcal{L}^2 := \mathcal{L}^2(\Omega, \mathcal{F}, \mathbf{P})$. Let $\mathcal{G}$ be a sub-σ-algebra of $\mathcal{F}$, and let $\mathcal{K} := \mathcal{L}^2(\mathcal{G}) := \mathcal{L}^2(\Omega, \mathcal{G}, \mathbf{P})$. By Section 6.10 applied to $\mathcal{G}$ rather than $\mathcal{F}$, we know that $\mathcal{K}$ is complete for the $\mathcal{L}^2$ norm. By Theorem 6.11 on orthogonal projection we know that there exists Y in $\mathcal{K} = \mathcal{L}^2(\mathcal{G})$ such that

(a) $$\mathbf{E}[(X - Y)^2] = \inf\{\mathbf{E}[(X - W)^2] : W \in \mathcal{L}^2(\mathcal{G})\},$$

(b) $$\langle X - Y, Z \rangle = 0, \quad \forall Z \text{ in } \mathcal{L}^2(\mathcal{G}).$$

Now, if $G \in \mathcal{G}$, then $Z := I_G \in \mathcal{L}^2(\mathcal{G})$ and (b) states that

$$\mathbf{E}(Y; G) = \mathbf{E}(X; G).$$

Hence Y *is* a version of $\mathbf{E}(X|\mathcal{G})$, as required.

Existence of $\mathbf{E}(X|\mathcal{G})$ for $X \in \mathcal{L}^1$

By splitting X as $X = X^+ - X_-$, we see that it is enough to deal with the case when $X \in (\mathcal{L}^1)^+$. So assume that $X \in (\mathcal{L}^1)^+$. We can now choose bounded variables X_n with $0 \leq X_n \uparrow X$. Since each X_n is in $\mathcal{L}^2$, we can choose a version Y_n of $\mathbf{E}(X_n|\mathcal{G})$. We now need to establish that

(c) *it is almost surely true that* $0 \leq Y_n \uparrow$.

We prove this in a moment. Given that (c) is true, we set

$$Y(\omega) := \limsup Y_n(\omega).$$

Then $Y \in m\mathcal{G}$, and $Y_n \uparrow Y$, a.s. But now (MON) allows us to deduce that

$$\mathbf{E}(Y; G) = \mathbf{E}(X; G) \qquad (G \in \mathcal{G})$$

from the corresponding result for Y_n and X_n.   □

A positivity result

Property (c) follows once we prove that

(d) *if U is a non-negative bounded RV, then*

$$E(U|\mathcal{G}) \geq 0, \quad \text{a.s.}$$

Proof of (d). Let W be a version of $E(U|\mathcal{G})$. If $P(W < 0) > 0$, then for some n, the set

$$G := \{W < -n^{-1}\} \text{ in } \mathcal{G} \text{ has positive probability,}$$

so that

$$0 \leq E(U; G) = E(W; G) < -n^{-1}P(G) < 0.$$

This contradiction finishes the proof. $\qquad\square$

9.6. Agreement with traditional usage

The case of two RVs will suffice to illustrate things. So suppose that X and Z are RVs which have a joint probability density function (pdf)

$$f_{X,Z}(x, z).$$

Then $f_Z(z) = \int_{\mathbf{R}} f_{X,Z}(x,z)dx$ acts as a probability density function for Z. Define the *elementary conditional pdf* $f_{X|Z}$ of X given Z via

$$f_{X|Z}(x|z) := \begin{cases} f_{X,Z}(x,z)/f_Z(z) & \text{if } f_Z(z) \neq 0; \\ 0 & \text{otherwise.} \end{cases}$$

Let h be a Borel function on $\mathbf{R}$ such that

$$E|h(X)| = \int_{\mathbf{R}} |h(x)| f_X(x)dx < \infty,$$

where of course $f_X(x) = \int_{\mathbf{R}} f_{X,Z}(x,z)dz$ gives a pdf for X. Set

$$g(z) := \int_{\mathbf{R}} h(x)f_{X|Z}(x|z)dx.$$

Then $Y := g(Z)$ *is a version of the conditional expectation of $h(X)$ given $\sigma(Z)$.*

Proof. The typical element of $\sigma(Z)$ has the form $\{\omega : Z(\omega) \in B\}$, where $B \in \mathcal{B}$. Hence, we must show that

(a) $\qquad L := E[h(X)I_B(Z)] = E[g(Z)I_B(Z)] =: R.$

But

$$L = \int\int h(x)I_B(z)f_{X,Z}(x,z)dxdz, \quad R = \int g(z)I_B(z)f_Z(z)dz,$$

and result (a) follows from Fubini's Theorem. $\qquad\square$

Some of the practice is given in Sections 15.6-15.9, which you can look at now.

▶▶▶**9.7. Properties of conditional expectation: a list**

These properties are proved in Section 9.8. All X's satisfy $E(|X|) < \infty$ in this list of properties. Of course, $\mathcal{G}$ and $\mathcal{H}$ denote sub-σ-algebras of $\mathcal{F}$. (The use of 'c' to denote 'conditional' in (cMON), etc., is obvious.)

(a) If Y is any version of $E(X|\mathcal{G})$ then $E(Y) = E(X)$. (*Very* useful, this.)

(b) If X is $\mathcal{G}$ measurable, then $E(X|\mathcal{G}) = X$, a.s.

(c) **(Linearity)** $E(a_1 X_1 + a_2 X_2|\mathcal{G}) = a_1 E(X_1|\mathcal{G}) + a_2 E(X_2|\mathcal{G})$, a.s.

Clarification: if Y_1 is a version of $E(X_1|\mathcal{G})$ and Y_2 is a version of $E(X_2|\mathcal{G})$, then $a_1 Y_1 + a_2 Y_2$ is a version of $E(a_1 X_1 + a_2 X_2|\mathcal{G})$.

(d) **(Positivity)** If $X \geq 0$, then $E(X|\mathcal{G}) \geq 0$, a.s.

(e) **(cMON)** If $0 \leq X_n \uparrow X$, then $E(X_n|\mathcal{G}) \uparrow E(X|\mathcal{G})$, a.s.

(f) **(cFATOU)** If $X_n \geq 0$, then $E[\liminf X_n|\mathcal{G}] \leq \liminf E[X_n|\mathcal{G}]$, a.s.

(g) **(cDOM)** If $|X_n(\omega)| \leq V(\omega)$, $\forall n$, $EV < \infty$, and $X_n \to X$, a.s., then

$$E(X_n|\mathcal{G}) \to E(X|\mathcal{G}), \quad \text{a.s.}$$

(h) **(cJENSEN)** If $c : \mathbf{R} \to \mathbf{R}$ is convex, and $E|c(X)| < \infty$, then

$$E[c(X)|\mathcal{G}] \geq c(E[X|\mathcal{G}]), \quad \text{a.s.}$$

Important corollary: $\|E(X|\mathcal{G})\|_p \leq \|X\|_p$ for $p \geq 1$.

(i) **(Tower Property)** If $\mathcal{H}$ is a sub-σ-algebra of $\mathcal{G}$, then

$$E[E(X|\mathcal{G})|\mathcal{H}] = E[X|\mathcal{H}], \quad \text{a.s.}$$

Note. We shorthand LHS to $E[X|\mathcal{G}|\mathcal{H}]$ for tidiness.

(j) **('Taking out what is known')** If Z is $\mathcal{G}$-measurable and bounded, then

(*) $$E[ZX|\mathcal{G}] = ZE[X|\mathcal{G}], \quad \text{a.s.}$$

If $p > 1$, $p^{-1} + q^{-1} = 1$, $X \in \mathcal{L}^p(\Omega, \mathcal{F}, \mathbf{P})$ and $Z \in \mathcal{L}^q(\Omega, \mathcal{G}, \mathbf{P})$, then (*) again holds. If $X \in (m\mathcal{F})^+$, $Z \in (m\mathcal{G})^+$, $E(X) < \infty$ and $E(ZX) < \infty$, then (*) holds.

(k) **(Rôle of independence)** If $\mathcal{H}$ is independent of $\sigma(\sigma(X), \mathcal{G})$, then

$$E[X|\sigma(\mathcal{G}, \mathcal{H})] = E(X|\mathcal{G}), \quad \text{a.s.}$$

In particular, if X is independent of $\mathcal{H}$, then $E(X|\mathcal{H}) = E(X)$, a.s.

9.8. Proofs of the properties in Section 9.7

Property (a) follows since $E(Y;\Omega) = E(X;\Omega)$, Ω being an element of $\mathcal{G}$. Property (b) is immediate from the definition, as is Property (c) now that its Clarification has been given.

Property (d) is not obvious, but the proof of (9.5,d) transfers immediately to our current situation.

Proof of (e). If $0 \le X_n \uparrow X$, then, by (d), if, for each n, Y_n is a version of $E(X_n|\mathcal{G})$, then (a.s.) $0 \le Y_n \uparrow$. Define $Y := \limsup Y_n$. Then $Y \in m\mathcal{G}$, and $Y_n \uparrow Y$, a.s. Now use (MON) to deduce from

$$E(Y_n; G) = E(X_n; G), \qquad \forall G \in \mathcal{G},$$

that $E(Y; G) = E(X; G)$, $\forall G \in \mathcal{G}$. (Of course we used a very similar argument in Section 9.5.) □

Proof of (f) *and* (g). You should check that the argument used to obtain (FATOU) from (MON) in Section 5.4 and the argument used to obtain (DOM) from (FATOU) in Section 5.9 both transfer without difficulty to yield the conditional versions. Doing the careful derivation of (cFATOU) from (cMON) and of (cDOM) from (cFATOU) is an essential exercise for you. □

Proof of (h). From (6.6,a), there exists a countable sequence $((a_n, b_n))$ of points in $\mathbf{R}^2$ such that

$$c(x) = \sup_n (a_n x + b_n), \quad x \in \mathbf{R}.$$

For each fixed n we deduce via (d) from $c(X) \ge a_n X + b_n$ that, almost surely,

$$(**) \qquad\qquad E[c(X)|\mathcal{G}] \ge a_n E[X|\mathcal{G}] + b_n.$$

By the usual appeal to countability, we can say that almost surely $(**)$ holds simultaneously for all n, whence, almost surely,

$$E[c(X)|\mathcal{G}] \ge \sup_n (a_n E[X|\mathcal{G}] + b_n) = c(E[X|\mathcal{G}]). \qquad □$$

Proof of corollary to (h). Let $p \ge 1$. Taking $c(x) = |x|^p$, we see that

$$E(|X|^p|\mathcal{G}) \ge |E(X|\mathcal{G})|^p, \text{ a.s.}$$

Now take expectations, using property (a).   □

Property (i) is virtually immediate from the definition of conditional expectation.

Proof of (j). Linearity shows that we can assume that $X \geq 0$. Fix a version Y of $E(X|\mathcal{G})$, and fix G in $\mathcal{G}$. We must prove that

> *if Z is $\mathcal{G}$-measurable and appropriate integrability conditions hold, then*

$$(***) \qquad\qquad E(ZX;G) = E(ZY;G).$$

We use the standard machine. If Z is the indicator of a set in $\mathcal{G}$, then $(***)$ is true by definition of the conditional expectation Y. Linearity then shows that $(***)$ holds for $Z \in SF^+(\Omega, \mathcal{G}, \mathbf{P})$. Next, (MON) shows that $(***)$ is true for $Z \in (m\mathcal{G})^+$ with the understanding that both sides might be infinite.

All that is necessary to establish that property (j) in the table is correct is to show that under each of the conditions given, $E(|ZX|) < \infty$. This is obvious if Z is bounded and X is in $\mathcal{L}^1$, and follows from the Hölder inequality if $X \in \mathcal{L}^p$ and $Z \in \mathcal{L}^q$ where $p > 1$ and $p^{-1} + q^{-1} = 1$.   □

Proof of (k). We can assume that $X \geq 0$ (and $E(X) < \infty$). For $G \in \mathcal{G}$ and $H \in \mathcal{H}$, XI_G and H are independent, so that by Theorem 7.1,

$$E(X; G \cap H) = E[(XI_G)I_H] = E(XI_G)\mathbf{P}(H).$$

Now if $Y = E(X|\mathcal{G})$ (a version of), then since Y is $\mathcal{G}$-measurable, YI_G is independent of $\mathcal{H}$ so that

$$E[(YI_G)I_H] = E(YI_G)\mathbf{P}(H)$$

and we have

$$E[X; G \cap H] = E[Y; G \cap H].$$

Thus the *measures*

$$F \mapsto E(X; F), \qquad F \mapsto E(Y; F)$$

on $\sigma(\mathcal{G}, \mathcal{H})$ of the same finite total mass agree on the π-system of sets of the form $G \cap H$ ($G \in \mathcal{G}, H \in \mathcal{H}$), and hence agree everywhere on $\sigma(\mathcal{G}, \mathcal{H})$. This is exactly what we had to prove.   □

9.9. Regular conditional probabilities and pdfs

For $F \in \mathcal{F}$, we have $\mathbb{P}(F) = \mathbb{E}(I_F)$. For $F \in \mathcal{F}$ and $\mathcal{G}$ a sub-σ-algebra of $\mathcal{F}$,

we define $\mathbb{P}(F|\mathcal{G})$ to be a version of $\mathbb{E}(I_F|\mathcal{G})$.

By linearity and (cMON), we can show that *for a fixed sequence (F_n) of disjoint elements of $\mathcal{F}$, we have*

(a) $$\mathbb{P}(\bigcup F_n|\mathcal{G}) = \sum \mathbb{P}(F_n|\mathcal{G}), \qquad \text{(a.s.)}$$

Except in trivial cases, there are *uncountably* many sequences of disjoint sets, so we cannot conclude from (a) that there exists a map

$$\mathbb{P}(\cdot,\cdot) : \Omega \times \mathcal{F} \to [0,1]$$

such that

(b1)   *for $F \in \mathcal{F}$, the function $\omega \mapsto \mathbb{P}(\omega, F)$ is a version of $\mathbb{P}(F|\mathcal{G})$;*

(b2)   *for almost every ω, the map*

$$F \mapsto \mathbb{P}(\omega, F)$$

is a probability measure on $\mathcal{F}$.

If such a map exists, it is called a *regular conditional probability* given $\mathcal{G}$. It is known that *regular conditional probabilities exist under most conditions encountered in practice*, but **they do not always exist**. The matter is too technical for a book at this level. See, for example, Parthasarathy (1967).

Important note. The elementary conditional pdf $f_{X|Z}(x|z)$ of Section 9.6 *is* a proper – technically, **regular** – **conditional pdf** for X given Z in that for every A in $\mathcal{B}$,

$$\omega \mapsto \int_A f_{X|Z}(x|Z(\omega))dx \text{ is a version of } \mathbb{P}(X \in A|Z).$$

Proof. Take $h = I_A$ in Section 9.6.   □

9.10. Conditioning under independence assumptions

Suppose that $r \in \mathbb{N}$ and that $X_1, X_2, \ldots, X_r$ are independent RVs, X_k having law Λ_k. If $h \in b\mathcal{B}^r$ and we define (for $x_1 \in \mathbb{R}$)

(a) $$\gamma^h(x_1) = \mathbb{E}[h(x_1, X_2, X_3, \ldots, X_r)],$$

then

(b) $\gamma^h(X_1)$ *is a version of the conditional expectation*
$$E[h(X_1, X_2, \ldots, X_r)|X_1].$$

Two proofs of (b). We need only show that for $B \in \mathcal{B}$,

(c) $$E[h(X_1, X_2, \ldots, X_r)I_B(X_1)] = E[\gamma^h(X_1)I_B(X_1)].$$

We can do this via the Monotone-Class Theorem, the class $\mathcal{H}$ of h satisfying (c) contains the indicator functions of elements in the π-system of sets of the form
$$B_1 \times B_2 \times \ldots \times B_r   (B_k \in \mathcal{B}),$$
etc., etc. Alternatively, we can appeal to the r-fold Fubini Theorem; for (c) says that

$$\int_{x \in \mathbf{R}^r} h(x)I_B(x_1)(\Lambda_1 \times \Lambda_2 \times \ldots \times \Lambda_r)(dx) = \int_{x_1 \in \mathbf{R}} \gamma^h(x_1)I_B(x_1)\Lambda_1(dx_1),$$

where

$$\gamma^h(x_1) = \int_{y \in \mathbf{R}^{r-1}} h(x_1, y)(\Lambda_2 \times \ldots \times \Lambda_r)(dy). \qquad \square$$

9.11. Use of symmetry: an example

Suppose that $X_1, X_2, \ldots$ are IID RVs with the same distribution as X, where $E(|X|) < \infty$. Let $S_n := X_1 + X_2 + \cdots + X_n$, and define
$$\mathcal{G}_n := \sigma(S_n, S_{n+1}, \ldots) = \sigma(S_n, X_{n+1}, X_{n+2}, \ldots).$$
We wish to calculate
$$E(X_1|\mathcal{G}_n),$$
for very good reasons, as we shall see in Chapter 14. Now $\sigma(X_{n+1}, X_{n+2}, \ldots)$ is independent of $\sigma(X_1, S_n)$ (which is a sub-σ-algebra of $\sigma(X_1, \ldots, X_n)$). Hence, by (9.7,k),
$$E(X_1|\mathcal{G}_n) = E(X_1|S_n).$$
But if Λ denotes the law of X, then, with s_n denoting $x_1 + x_2 + \cdots + x_n$, we have
$$E(X_1; S_n \in B)$$
$$= \int \cdots \int_{s_n \in B} x_1 \Lambda(dx_1)\Lambda(dx_2)\ldots\Lambda(dx_n)$$
$$= E(X_2; S_n \in B) = \cdots = E(X_n; S_n \in B).$$
Hence, almost surely,
$$E(X_1|S_n) = \cdots = E(X_n|S_n)$$
$$= n^{-1}E(X_1 + \cdots + X_n|S_n) = n^{-1}S_n.$$

Chapter 10
Martingales

10.1. Filtered spaces

►►As basic datum, we now take a *filtered space* $(\Omega, \mathcal{F}, \{\mathcal{F}_n\}, \mathbf{P})$. Here,

$(\Omega, \mathcal{F}, \mathbf{P})$ is a probability triple as usual,

$\{\mathcal{F}_n : n \geq 0\}$ is a **filtration**,   that is, an increasing family of sub-σ-algebras of $\mathcal{F}$:

$$\mathcal{F}_0 \subseteq \mathcal{F}_1 \subseteq \ldots \subseteq \mathcal{F}.$$

We define

$$\mathcal{F}_\infty := \sigma\left(\bigcup_n \mathcal{F}_n\right) \subseteq \mathcal{F}.$$

Intuitive idea. The information about ω in Ω available to us at (or, if you prefer, 'just after') time n consists precisely of the values of $Z(\omega)$ for all $\mathcal{F}_n$ measurable functions Z. Usually, $\{\mathcal{F}_n\}$ is the **natural filtration**

$$\mathcal{F}_n = \sigma(W_0, W_1, \ldots, W_n)$$

of some (stochastic) process $W = (W_n : n \in \mathbf{Z}^+)$, and then the information about ω which we have at time n consists of the values

$$W_0(\omega), W_1(\omega), \ldots, W_n(\omega).$$

10.2. Adapted process

►A process $X = (X_n : n \geq 0)$ is called *adapted* (to the filtration $\{\mathcal{F}_n\}$) if for each n, X_n is $\mathcal{F}_n$-measurable.

Intuitive idea. If X is adapted, the value $X_n(\omega)$ is known to us at time n. Usually, $\mathcal{F}_n = \sigma(W_0, W_1, \ldots, W_n)$ and $X_n = f_n(W_0, W_1, \ldots, W_n)$ for some $\mathcal{B}^{n+1}$-measurable function f_n on $\mathbf{R}^{n+1}$.

10.3. Martingale, supermartingale, submartingale

▶▶▶A process X is called a **martingale** (relative to $(\{\mathcal{F}_n\}, \mathbf{P})$) if

(i) X is adapted,

(ii) $\mathbf{E}(|X_n|) < \infty,\ \forall n,$

(iii) $\mathbf{E}[X_n|\mathcal{F}_{n-1}] = X_{n-1},\quad$ a.s.$\quad\quad (n \geq 1).$

A **supermartingale** (relative to $(\{\mathcal{F}_n\}, \mathbf{P})$) is defined similarly, except that (iii) is replaced by

$$\mathbf{E}[X_n|\mathcal{F}_{n-1}] \leq X_{n-1},\quad \text{a.s.}\quad\quad (n \geq 1),$$

and a **submartingale** is defined with (iii) replaced by

$$\mathbf{E}[X_n|\mathcal{F}_{n-1}] \geq X_{n-1},\quad \text{a.s.}\quad\quad (n \geq 1).$$

A **super**martingale 'decreases on average'; a **sub**martingale 'increases on average'! [Supermartingale corresponds to superharmonic: a function f on $\mathbf{R}^n$ is superharmonic if and only if for a Brownian motion B on $\mathbf{R}^n$, $f(B)$ is a local supermartingale relative to the natural filtration of B. Compare Section 10.13.]

Note that X is a supermartingale if and only if $-X$ is a submartingale, and that X is a martingale if and only if it is both a supermartingale and a submartingale. It is important to note that a process X for which $X_0 \in \mathcal{L}^1(\Omega, \mathcal{F}_0, \mathbf{P})$ is a martingale [respectively, supermartingale, submartingale] if and only if the process $X - X_0 = (X_n - X_0 : n \in \mathbf{Z}^+)$ has the same property. So we can focus attention on processes which are null at 0.

▶   If X is for example a supermartingale, then the Tower Property of CEs, (9.7)(i), shows that for $m < n$,

$$\mathbf{E}[X_n|\mathcal{F}_m] = \mathbf{E}[X_n|\mathcal{F}_{n-1}|\mathcal{F}_m] \leq \mathbf{E}[X_{n-1}|\mathcal{F}_m] \leq \ldots$$
$$\leq X_m,\quad \text{a.s..}$$

10.4. Some examples of martingales

As we shall see, it is very helpful to view all martingales, supermartingales and submartingales in terms of gambling. But, of course, the enormous importance of martingale theory derives from the fact that martingales crop up in very many contexts. For example, diffusion theory, which used to be studied via methods from Markov-process theory, from the theory of partial differential equations, etc., has been revolutionized by the martingale approach.

Let us now look at some simple first examples, and mention an interesting question (solved later) pertaining to each.

(a) **Sums of independent zero-mean RVs.** Let $X_1, X_2, \ldots$ be a sequence of *independent* RVs with $E(|X_k|) < \infty$, $\forall k$, and

$$E(X_k) = 0, \quad \forall k.$$

Define ($S_0 := 0$ and)

$$S_n := X_1 + X_2 + \cdots + X_n,$$
$$\mathcal{F}_n := \sigma(X_1, X_2, \ldots, X_n), \qquad \mathcal{F}_0 := \{\emptyset, \Omega\}.$$

Then for $n \geq 1$, we have (a.s.)

$$E(S_n | \mathcal{F}_{n-1}) \stackrel{(c)}{=} E(S_{n-1} | \mathcal{F}_{n-1}) + E(X_n | \mathcal{F}_{n-1})$$
$$\stackrel{(b,k)}{=} S_{n-1} + E(X_n) = S_{n-1}.$$

The first (a.s.) equality is obvious from the linearity property (9.7,c). Since S_{n-1} is $\mathcal{F}_{n-1}$-measurable, we have $E(S_{n-1} | \mathcal{F}_{n-1}) = S_{n-1}$ (a.s.) by (9.7,b); and since X_n is independent of $\mathcal{F}_{n-1}$, we have $E(X_n | \mathcal{F}_{n-1}) = E(X_n)$ (a.s.) by (9.7,k). That must explain our notation!

Interesting question: when does $\lim S_n$ exist (a.s.)? See Section 12.5.

(b) **Products of non-negative independent RVs of mean 1.** Let $X_1, X_2, \ldots$ be a sequence of independent non-negative random variables with

$$E(X_k) = 1, \quad \forall k.$$

Define ($M_0 := 1$, $\mathcal{F}_0 := \{\emptyset, \Omega\}$ and)

$$M_n := X_1 X_2 \ldots X_n, \qquad \mathcal{F}_n := \sigma(X_1, X_2, \ldots, X_n).$$

Then, for $n \geq 1$, we have (a.s.)

$$E(M_n | \mathcal{F}_{n-1}) = E(M_{n-1} X_n | \mathcal{F}_{n-1}) \stackrel{(j)}{=} M_{n-1} E(X_n | \mathcal{F}_{n-1})$$
$$\stackrel{(k)}{=} M_{n-1} E(X_n) = M_{n-1},$$

so that M *is a martingale.*

It should be remarked that such martingales are not at all artificial.

Interesting question. Because M is a non-negative martingale, $M_\infty = \lim M_n$ exists (a.s.); this is part of the Martingale Convergence Theorem of the next chapter. When can we say that $E(M_\infty) = 1$? See Sections 14.12 and 14.17.

(c) **Accumulating data about a random variable.** Let $\{\mathcal{F}_n\}$ be our filtration, and let $\xi \in \mathcal{L}^1(\Omega, \mathcal{F}, \mathbf{P})$. Define $M_n := E(\xi|\mathcal{F}_n)$ ('some version of'). By the Tower Property (9.7,i), we have (a.s.)

$$E(M_n|\mathcal{F}_{n-1}) = E(\xi|\mathcal{F}_n|\mathcal{F}_{n-1}) = E(\xi|\mathcal{F}_{n-1}) = M_{n-1}.$$

Hence M is a martingale.

Interesting question. In this case, we shall be able to say that

$$M_n \to M_\infty := E(\xi|\mathcal{F}_\infty), \quad \text{a.s.,}$$

because of Lévy's Upward Theorem (Chapter 14). Now M_n is the best predictor of ξ given the information available to us at time n, and M_∞ is the best prediction of ξ we can ever make. When can we say that $\xi = E(\xi|\mathcal{F}_\infty)$, a.s? The answer is not always obvious. See Section 15.8.

10.5. Fair and unfair games

Think now of

$X_n - X_{n-1}$ as your *net winnings per unit stake* in game n $(n \geq 1)$

in a series of games, played at times $n = 1, 2, \ldots$. *There is no game at time 0.*

In the martingale case,

(a) $E[X_n - X_{n-1}|\mathcal{F}_{n-1}] = 0,$ (game series is fair),

and in the supermartingale case,

(b) $E[X_n - X_{n-1}|\mathcal{F}_{n-1}] \leq 0,$ (game series is unfavourable to you).

Note that (a) [respectively (b)] gives a useful way of formulating the martingale [supermartingale] property of X.

10.6. Previsible process, gambling strategy

▶▶We call a process $C = (C_n : n \in \mathbf{N})$ **previsible** if

C_n is $\mathcal{F}_{n-1}$ measurable $(n \geq 1)$.

Note that C has parameter set N rather than Z^+: C_0 does not exist.

Think of C_n as your stake on game n. You have to decide on the value of C_n based on the history up to (and including) time $n-1$. This is the intuitive significance of the 'previsible' character of C. Your winnings on game n are $C_n(X_n - X_{n-1})$ and *your total winnings up to time n are*

$$Y_n = \sum_{1 \leq k \leq n} C_k(X_k - X_{k-1}) =: (C \bullet X)_n.$$

Note that $(C \bullet X)_0 = 0$, and that

$$Y_n - Y_{n-1} = C_n(X_n - X_{n-1}).$$

The expression $C \bullet X$, the **martingale transform** of X by C, is the discrete analogue of the **stochastic integral** $\int C\,dX$. Stochastic-integral theory is one of the greatest achievements of the modern theory of probability.

10.7. A fundamental principle: you can't beat the system!

▶▶(i)   *Let C be a bounded non-negative previsible process so that, for some K in $[0, \infty)$, $|C_n(\omega)| \leq K$ for every n and every ω. Let X be a supermartingale [respectively martingale]. Then $C \bullet X$ is a supermartingale [martingale] null at 0.*

(ii)   *If C is a bounded previsible process and X is a martingale, then $(C \bullet X)$ is a martingale null at 0.*

(iii)   *In (i) and (ii), the boundedness condition on C may be replaced by the condition $C_n \in \mathcal{L}^2, \forall n$, provided we also insist that $X_n \in \mathcal{L}^2, \forall n$.*

Proof of (i). Write Y for $C \bullet X$. Since C_n is bounded non-negative and $\mathcal{F}_{n-1}$ measurable,

$$\mathsf{E}[Y_n - Y_{n-1}|\mathcal{F}_{n-1}] \overset{(j)}{=} C_n \mathsf{E}[X_n - X_{n-1}|\mathcal{F}_{n-1}] \leq 0, \ [\text{resp.} =0].$$

Proofs of (ii) and (iii) are now obvious. (Look again at (9.7,j).) $\square$

10.8. Stopping time

A map $T : \Omega \to \{0, 1, 2, \ldots; \infty\}$ is called a *stopping time* if,

▶▶(a) $$\{T \leq n\} = \{\omega : T(\omega) \leq n\} \in \mathcal{F}_n, \quad \forall n \leq \infty,$$

equivalently,

(b) $\{T = n\} = \{\omega : T(\omega) = n\} \in \mathcal{F}_n, \quad \forall n \leq \infty.$

Note that T can be ∞.

Proof of the equivalence of (a) *and* (b). If T has property (a), then

$$\{T = n\} = \{T \leq n\} \backslash \{T \leq n - 1\} \in \mathcal{F}_n.$$

If T has property (b), then for $k \leq n$, $\{T = k\} \in \mathcal{F}_k \subseteq \mathcal{F}_n$ and

$$\{T \leq n\} = \bigcup_{0 \leq k \leq n} \{T = k\} \in \mathcal{F}_n.$$

Intuitive idea. T is a time when you can decide to stop playing our game. Whether or not you stop immediately after the nth game depends only on the history up to (and including) time $n : \{T = n\} \in \mathcal{F}_n.$

Example. Suppose that (A_n) is an adapted process, and that $B \in \mathcal{B}$. Let

$$T = \inf\{n \geq 0 : A_n \in B\} = \text{time of first entry of } A \text{ into set } B.$$

By convention, $\inf(\emptyset) = \infty$, so that $T = \infty$ if A never enters set B. Obviously,

$$\{T \leq n\} = \bigcup_{k \leq n} \{A_k \in B\} \in \mathcal{F}_n,$$

so that T is a stopping time.

Example. Let $L = \sup\{n : n \leq 10; A_n \in B\}$, $\sup(\emptyset) = 0$. Convince yourself that L is *NOT* a stopping time (unless A is freaky).

10.9. Stopped supermartingales are supermartingales

Let X be a supermartingale, and let T be a stopping time.

Suppose that you always bet 1 unit and quit playing at (immediately after) time T. Then your 'stake process' is $C^{(T)}$, where, for $n \in \mathbb{N}$,

$$C_n^{(T)} = I_{\{n \leq T\}}, \quad \text{so} \quad C_n^{(T)}(\omega) = \begin{cases} 1 & \text{if } n \leq T(\omega), \\ 0 & \text{otherwise.} \end{cases}$$

Your 'winnings process' is the process with value at time n equal to

$$(C^{(T)} \bullet X)_n = X_{T \wedge n} - X_0.$$

If X^T denotes the process X **stopped at** T:

$$X_n^T(\omega) := X_{T(\omega) \wedge n}(\omega),$$

then

$$C^{(T)} \bullet X = X^T - X_0.$$

Now $C^{(T)}$ is clearly bounded (by 1) and non-negative. Moreover, $C^{(T)}$ is previsible because $C_n^{(T)}$ can only be 0 or 1 and, for $n \in \mathbb{N}$,

$$\{C_n^{(T)} = 0\} = \{T \leq n - 1\} \in \mathcal{F}_{n-1}.$$

Result 10.7 now yields the following result.

THEOREM

▶▶(i)   *If X is a supermartingale and T is a stopping time, then the* **stopped process** $X^T = (X_{T \wedge n} : n \in \mathbb{Z}^+)$ *is a supermartingale, so that in particular,*

$$\mathbb{E}(X_{T \wedge n}) \leq \mathbb{E}(X_0), \quad \forall n.$$

▶▶(ii)   *If X is a martingale and T is a stopping time, then X^T is a martingale, so that in particular,*

$$\mathbb{E}(X_{T \wedge n}) = \mathbb{E}(X_0), \quad \forall n.$$

It is important to notice that this theorem imposes no extra integrability conditions whatsoever (except of course for those implicit in the definition of supermartingale and martingale).

But be very careful! Let X be a simple random walk on $\mathbb{Z}$, starting at 0. Then X is a martingale. Let T be the stopping time:

$$T := \inf\{n : X_n = 1\}.$$

It is well known that $\mathbb{P}(T < \infty) = 1$. (See Section 10.12 for a martingale proof of this fact, and for a martingale calculation of the distribution of T.) However, even though

$$\mathbb{E}(X_{T \wedge n}) = \mathbb{E}(X_0) \text{ for every } n,$$

we have

$$1 = \mathbb{E}(X_T) \neq \mathbb{E}(X_0) = 0.$$

We very much want to know when we *can* say that

$$E(X_T) = E(X_0)$$

for a martingale X. The following theorem gives some sufficient conditions.

10.10. Doob's Optional-Stopping Theorem

▶▶(a)   *Let T be a stopping time. Let X be a supermartingale. Then X_T is integrable and*

$$E(X_T) \leq E(X_0)$$

in each of the following situations:

(i) *T is bounded (for some N in N, $T(\omega) \leq N$, $\forall \omega$);*

(ii) *X is bounded (for some K in R^+, $|X_n(\omega)| \leq K$ for every n and every ω) and T is a.s. finite;*

(iii) *$E(T) < \infty$, and, for some K in R^+,*

$$|X_n(\omega) - X_{n-1}(\omega)| \leq K \quad \forall(n,\omega).$$

(b)   *If any of the conditions (i)-(iii) holds and X is a martingale, then*

$$E(X_T) = E(X_0).$$

Proof of (a). We know that $X_{T \wedge n}$ is integrable, and

(∗) $$E(X_{T \wedge n} - X_0) \leq 0.$$

For (i), we can take $n = N$. For (ii), we can let $n \to \infty$ in (∗) using (BDD). For (iii), we have

$$|X_{T \wedge n} - X_0| = |\sum_{k=1}^{T \wedge n}(X_k - X_{k-1})| \leq KT$$

and $E(KT) < \infty$, so that (DOM) justifies letting $n \to \infty$ in (∗) to obtain the answer we want. □

Proof of (b). Apply (a) to X and to $(-X)$. □

Corollary

►►(c)   *Suppose that M is a martingale, the increments $M_n - M_{n-1}$ of which are bounded by some constant K_1. Suppose that C is a previsible process bounded by some constant K_2, and that T is a stopping time such that $E(T) < \infty$. Then*

$$E(C \bullet M)_T = 0.$$

Proof of the following final part of the Optional-Stopping Theorem is left as an **Exercise**. (It's clear whose lemma is needed!)

(d)   *If X is a non-negative supermartingale, and T is a stopping time which is a.s. finite, then*

$$E(X_T) \leq E(X_0).$$

10.11. Awaiting the almost inevitable

In order to be able to apply some of the results of the preceding Section, we need ways of proving that (when true!) $E(T) < \infty$. The following announcement of the principle that *'whatever always stands a reasonable chance of happening will almost surely happen – sooner rather than later'* is often useful.

LEMMA

►   *Suppose that T is a stopping time such that for some N in $\mathbb{N}$ and some $\varepsilon > 0$, we have, for every n in $\mathbb{N}$:*

$$P(T \leq n + N | \mathcal{F}_n) > \varepsilon, \quad \text{a.s.}$$

Then $E(T) < \infty$.

You will find the proof of this set as an exercise in Chapter E.

Note that if T is the first occasion by which the monkey in the 'Tricky exercise' at the end of Section 4.9 first completes

ABRACADABRA,

then $E(T) < \infty$. You will find another exercise in Chapter E inviting you to apply result (c) of the preceding Section to show that

$$E(T) = 26^{11} + 26^4 + 26.$$

A large number of other Exercises are now accessible to you.

10.12. Hitting times for simple random walk

Suppose that $(X_n : n \in \mathbf{N})$ is a sequence of IID RVs, each X_n having the same distribution as X where

$$\mathbf{P}(X = 1) = \mathbf{P}(X = -1) = \tfrac{1}{2}.$$

Define $S_0 := 0$, $S_n := X_1 + \cdots + X_n$, and set

$$T := \inf\{n : S_n = 1\}.$$

Let

$$\mathcal{F}_n = \sigma(X_1, \ldots, X_n) = \sigma(S_0, S_1, \ldots, S_n).$$

Then the process S is adapted (to $\{\mathcal{F}_n\}$), so that T is a stopping time. We wish to calculate the distribution of T.

For $\theta \in \mathbf{R}$, $\mathbf{E}e^{\theta X} = \tfrac{1}{2}(e^\theta + e^{-\theta}) = \cosh\theta$, so that

$$\mathbf{E}[(\operatorname{sech}\theta)e^{\theta X_n}] = 1, \quad \forall n.$$

Example (10.4,b) shows that M^θ *is a martingale,* where

$$M_n^\theta = (\operatorname{sech}\theta)^n e^{\theta S_n}.$$

Since T is a stopping time, and M^θ is a martingale, we have

(a) $\qquad\qquad \mathbf{E}M_{T \wedge n}^\theta = \mathbf{E}[(\operatorname{sech}\theta)^{T \wedge n} \exp(\theta S_{T \wedge n})] = 1, \quad \forall n.$

▶   *Now insist that $\theta > 0$.*

Then, firstly, $\exp(\theta S_{T \wedge n})$ is bounded by e^θ, so $M_{T \wedge n}^\theta$ is bounded by e^θ. Secondly, as $n \uparrow \infty$, $M_{T \wedge n}^\theta \to M_T^\theta$ *where the latter is defined to be 0 if $T = \infty$.* The Bounded Convergence Theorem allows us to let $n \to \infty$ in (a) to obtain

$$\mathbf{E}M_T^\theta = 1 = \mathbf{E}[(\operatorname{sech}\theta)^T e^\theta]$$

the term inside $[\cdot]$ on the right-hand side correctly being 0 if $T = \infty$. Hence

(b) $\qquad\qquad \mathbf{E}[(\operatorname{sech}\theta)^T] = e^{-\theta} \quad \text{for } \theta > 0.$

We now let $\theta \downarrow 0$. Then $(\operatorname{sech}\theta)^T \uparrow 1$ if $T < \infty$, and $(\operatorname{sech}\theta)^T \uparrow 0$ if $T = \infty$. Either (MON) or (BDD) yields

$$\mathbf{E}I_{\{T < \infty\}} = 1 = \mathbf{P}(T < \infty).$$

▶   The above argument has been given carefully to show how to deal with possibly infinite stopping times.

Put $\alpha = \operatorname{sech}\theta$ in (b) to obtain

(c) $\mathsf{E}(\alpha^T) = \sum \alpha^n \mathsf{P}(T = n) = e^{-\theta} = \alpha^{-1}[1 - \sqrt{1 - \alpha^2}],$

so that

$$\mathsf{P}(T = 2m - 1) = (-1)^{m+1}\binom{\frac{1}{2}}{m}.$$

Intuitive proof of (c)

We have

(d) $\begin{aligned} f(\alpha) :&= \mathsf{E}(\alpha^T) = \tfrac{1}{2}\mathsf{E}(\alpha^T|X_1 = 1) + \tfrac{1}{2}\mathsf{E}(\alpha^T|X_1 = -1) \\ &= \tfrac{1}{2}\alpha + \tfrac{1}{2}\alpha f(\alpha)^2. \end{aligned}$

The intuitive reason for the very last term is that time 1 has already elapsed giving the α, and the time taken to go from -1 to 1 has the form $T_1 + T_2$, where T_1 (the time to go from -1 to 0) and T_2 (the time to go from 0 to 1) are independent, each with the same distribution as T. It is *not obvious* that 'T_1 and T_2 are independent', but it is not difficult to devise a proof: the so-called *Strong Markov Theorem* would allow us to justify (d).

10.13. Non-negative superharmonic functions for Markov chains

Let E be a finite or countable set. Let $P = (p_{ij})$ be a stochastic $E \times E$ matrix, so that, for $i, j \in E$, we have

$$p_{ij} \ge 0, \qquad \sum_{k \in E} p_{ik} = 1.$$

Let μ be a probability measure on E. We know from Section 4.8 that there exists a triple $(\Omega, \mathcal{F}, \mathsf{P}^\mu)$ (we now signify the dependence of P on μ) carrying a Markov chain $Z = (Z_n : n \in \mathbf{Z}^+)$ such that (4.8,a) holds. We write 'a.s., P^μ' to signify 'almost surely relative to the P^μ-measure'.

Let $\mathcal{F}_n := \sigma(Z_0, Z_1, \ldots, Z_n)$. It is easy to deduce from (4.8,a) that if we write $p(i, j)$ instead of p_{ij} when typographically convenient, then (a.s.,P^μ)

$$\mathsf{P}^\mu(Z_{n+1} = j|\mathcal{F}_n) = p(Z_n, j).$$

Let h be a non-negative function on E and define the function Ph on E via

$$(Ph)(i) = \sum_j p(i, j)h(j).$$

Assume that our non-negative function h is finite and *P-superharmonic* in that $Ph \leq h$ on E. Then, (cMON) shows that, a.s., $\mathbf{P}^\mu$,

$$\mathbf{E}^\mu[h(Z_{n+1})|\mathcal{F}_n] = \sum p(Z_n, j)h(j) = (Ph)(Z_n) \leq h(Z_n),$$

so that $h(Z_n)$ *is a non-negative supermartingale* (whatever be the initial distribution μ).

Suppose that the chain Z is *irreducible recurrent* in that

$$f_{ij} := \mathbf{P}^i(T_j < \infty) = 1, \quad \forall i, j \in E,$$

where $\mathbf{P}^i$ denotes $\mathbf{P}^\mu$ when μ is the unit mass ($\mu_j = \delta_{ij}$) at i (see 'Note' below) and

$$T_j := \inf\{n : n \geq 1; Z_n = j\}.$$

Note that the infimum is over $\{n \geq 1\}$, so that f_{ii} is the probability of a *return* to i if Z starts at i. Then, by Theorem 10.10(d), we see that *if h is non-negative and P-superharmonic*, then, for any i and j in E,

$$h(j) = \mathbf{E}^i h(Z_{T_j}) \leq \mathbf{E}^i h(Z_0) = h(i),$$

so that h *is constant on E.*

Exercise. Explain (at first intuitively, and later with consideration of rigour) why

$$f_{ij} = \sum_{k \neq j} p_{ik} f_{kj} + p_{ij} \geq \sum_k p_{ik} f_{kj}$$

and deduce that if every non-negative P-superharmonic function is constant, then Z is irreducible recurrent.   □

So we have proved that

> *our chain Z is irreducible and recurrent if and only if every non-negative P-superharmonic function is constant.*

This is a trivial first step in the links between probability and potential theory.

Note. The perspicacious reader will have been upset by a lack of precision in this section. I wished to convey what is interesting first.

Only the *very* enthusiastic should read the remainder of this section.

The natural thing to do, given the one-step transition matrix P, is to take the *canonical model* for the Markov chain Z obtained as follows. Let $\mathcal{E}$ denote the σ-algebra of all subsets of E and define

$$(\Omega, \mathcal{F}) := \prod_{n \in \mathbf{Z}^+} (E, \mathcal{E}).$$

In particular, a point ω of Ω is a sequence

$$\omega = (\omega_0, \omega_1, \ldots)$$

of elements of E. For ω in Ω and n in $\mathbf{Z}^+$, define

$$Z_n(\omega) := \omega_n \in E.$$

Then, for each probability measure μ on $(E, \mathcal{E})$, there is a unique probability measure $\mathbf{P}^\mu$ on $(\Omega, \mathcal{F})$ such that for $n \in \mathbf{N}$ and $i_0, i_1, \ldots, i_n \in E$, we have

$$(*) \quad \mathbf{P}^\mu[\omega : Z_0(\omega) = i_0, Z_1(\omega) = i_1, \ldots, Z_n(\omega) = i_n] = \mu_{i_0} p_{i_0 i_1} \cdots p_{i_{n-1} i_n}.$$

The uniqueness is trivial because ω-sets of the form contained in $[\cdot]$ on the left-hand side of $(*)$, together with $\emptyset$, form a π-system generating $\mathcal{F}$. Existence follows because we can take $\mathbf{P}^\mu$ to be the $\tilde{\mathbf{P}}^\mu$-law of the non-canonical process $\tilde{Z}$ constructed in Section A4.3:

$$\mathbf{P}^\mu = \tilde{\mathbf{P}}^\mu \circ \tilde{Z}^{-1}.$$

Here, we regard $\tilde{Z}$ as the map

$$\tilde{Z} : \tilde{\Omega} \to \Omega$$
$$\tilde{\omega} \mapsto (\tilde{Z}_0(\tilde{\omega}), \tilde{Z}_1(\tilde{\omega}), \ldots),$$

this map $\tilde{Z}$ being $\tilde{\mathcal{F}}/\mathcal{F}$ measurable in that

$$Z^{-1} : \mathcal{F} \to \tilde{\mathcal{F}}.$$

The canonical model thus obtained is very satisfying because the measurable space $(\Omega, \mathcal{F})$ carries all measures $\mathbf{P}^\mu$ simultaneously.

Chapter 11
The Convergence Theorem

11.1. The picture that says it all

The top part of Figure 11.1 shows a sample path $n \mapsto X_n(\omega)$ for a process X where $X_n - X_{n-1}$ represents your winnings per unit stake on game n. The lower part of the picture illustrates your total-winnings process $Y := C \bullet X$ under the previsible strategy C described as follows:

> Pick two numbers a and b with $a < b$.
> REPEAT
>> Wait until X gets below a
>> Play unit stakes until X gets above b and stop playing
> UNTIL FALSE (that is, forever!).

Black blobs signify where $C = 1$; and open circles signify where $C = 0$. Recall that C is not defined at time 0.

To be more formal (and to prove inductively that C is previsible), define

$$C_1 := I_{\{X_0 < a\}},$$

and, for $n \geq 2$,

$$C_n := I_{\{C_{n-1}=1\}} I_{\{X_{n-1} \leq b\}} + I_{\{C_{n-1}=0\}} I_{\{X_{n-1} < a\}}.$$

11.2. Upcrossings

The **number $U_N[a,b](\omega)$ of upcrossings of $[a,b]$ made by** $n \mapsto X_n(\omega)$ **by time N** is defined to be the largest k in $\mathbf{Z}^+$ such that we can find

$$0 \leq s_1 < t_1 < s_2 < t_2 < \cdots < s_k < t_k \leq N$$

with

$$X_{s_i}(\omega) < a, \quad X_{t_i}(\omega) > b \qquad (1 \leq i \leq k).$$

Figure 11.1

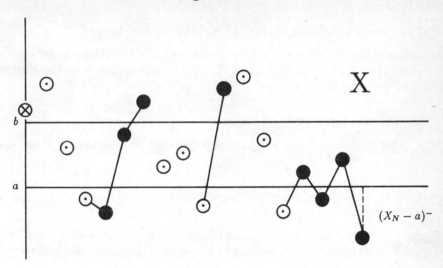

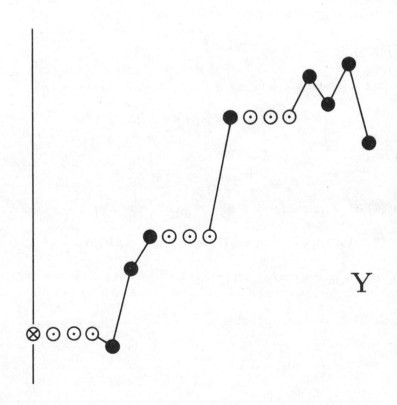

The fundamental inequality (recall that $Y_0(\omega) := 0$)

►(D) $$Y_N(\omega) \geq (b-a)U_N[a,b](\omega) - [X_N(\omega) - a]^-$$

is obvious from the picture: every upcrossing of $[a,b]$ increases the Y-value by at least $(b-a)$, while the $[X_N(\omega) - a]^-$ overemphasizes the loss during the last 'interval of play'.

11.3. Doob's Upcrossing Lemma

►   *Let X be a supermartingale. Let $U_N[a,b]$ be the number of upcrossings of $[a,b]$ by time N. Then*

$$(b-a)\mathsf{E}U_N[a,b] \leq \mathsf{E}[(X_N - a)^-].$$

Proof. The process C is previsible, bounded and ≥ 0, and $Y = C \bullet X$. Hence Y is a supermartingale, and $\mathsf{E}(Y_N) \leq 0$. The result now follows from (11.2,D).

11.4. COROLLARY

►   *Let X be a supermartingale bounded in $\mathcal{L}^1$ in that*

$$\sup_n \mathsf{E}(|X_n|) < \infty.$$

Let $a, b \in \mathbf{R}$ with $a < b$. Then, with $U_\infty[a,b] :=\uparrow \lim_N U_N[a,b]$,

$$(b-a)\mathsf{E}U_\infty[a,b] \leq |a| + \sup_n \mathsf{E}(|X_n|) < \infty$$

so that

$$\mathsf{P}(U_\infty[a,b] = \infty) = 0.$$

Proof. By (11.3), we have, for $N \in \mathbf{N}$,

$$(b-a)\mathsf{E}U_N[a,b] \leq |a| + \mathsf{E}(|X_N|) \leq |a| + \sup_n \mathsf{E}(|X_n|).$$

Now let $N \uparrow \infty$, using (MON).   □

11.5. Doob's 'Forward' Convergence Theorem

▶▶   **Let X be a supermartingale bounded in $\mathcal{L}^1$: $\sup\limits_n E(|X_n|) < \infty$.**

Then, almost surely, $X_\infty := \lim X_n$ exists and is finite. For definiteness, we *define* $X_\infty(\omega) := \limsup X_n(\omega)$, $\forall\omega$, so that X_∞ is $\mathcal{F}_\infty$ measurable and $X_\infty = \lim X_n$, a.s.

Proof (Doob). Write (noting the use of $[-\infty, \infty]$):

$$
\begin{aligned}
\Lambda :&= \{\omega : X_n(\omega) \quad \text{does not converge to a limit in} \quad [-\infty, \infty]\} \\
&= \{\omega : \liminf X_n(\omega) < \limsup X_n(\omega)\} \\
&= \bigcup_{\{a,b\in\mathbf{Q}:a<b\}} \{\omega : \liminf X_n(\omega) < a < b < \limsup X_n(\omega)\} \\
&=: \bigcup \Lambda_{a,b} \quad \text{(say)}.
\end{aligned}
$$

But

$$
\Lambda_{a,b} \subseteq \{\omega : U_\infty[a, b](\omega) = \infty\},
$$

so that, by (11.4), $P(\Lambda_{a,b}) = 0$. Since Λ is a countable union of sets $\Lambda_{a,b}$, we see that $P(\Lambda) = 0$, whence

$$
X_\infty := \lim X_n \text{ exists a.s. in } [-\infty, \infty].
$$

But Fatou's Lemma shows that

$$
\begin{aligned}
E(|X_\infty|) = E(\liminf |X_n|) &\leq \liminf E(|X_n|) \\
&\leq \sup E(|X_n|) < \infty,
\end{aligned}
$$

so that

$$
P(X_\infty \text{ is finite}) = 1. \qquad \square
$$

Note. There are other proofs for the discrete-parameter case. None of these is as probabilistic, and none shares the central importance of this one for the continuous-parameter case.

11.6. Warning

As we saw for the branching-process example, it need not be true that $X_n \to X_\infty$ in $\mathcal{L}^1$.

11.7. Corollary

▶▶   **If X is a non-negative supermartingale, then $X_\infty := \lim X_n$ exists almost surely.**

Proof. X is obviously bounded in $\mathcal{L}^1$, since $E(|X_n|) = E(X_n) \leq E(X_0)$. $\square$

Chapter 12

Martingales bounded in $\mathcal{L}^2$

12.0. Introduction

When it works, one of the easiest ways of proving that a martingale M is bounded in $\mathcal{L}^1$ is to prove that it is *bounded in $\mathcal{L}^2$* in the sense that

(a) $$\sup_n \|M_n\|_2 < \infty, \quad \text{equivalently,} \quad \sup_n \mathsf{E}(M_n^2) < \infty.$$

Boundedness in $\mathcal{L}^2$ is often easy to check because of a Pythagorean formula (proved in Section 12.1)

$$\mathsf{E}(M_n^2) = \mathsf{E}(M_0^2) + \sum_{k=1}^n \mathsf{E}[(M_k - M_{k-1})^2].$$

The study of sums of independent random variables, a central topic in the classical theory, will be seen to hinge on Theorem 12.2 below, both parts of which have neat martingale proofs. We shall prove the **Three-Series Theorem**, which says exactly when a sum of independent random variables converges. We shall also prove the general **Strong Law of Large Numbers** for IID RVs and **Lévy's extension of the Borel-Cantelli Lemmas**.

12.1. Martingales in $\mathcal{L}^2$; orthogonality of increments

Let $M = (M_n : n \geq 0)$ be a *martingale in $\mathcal{L}^2$* in that each M_n is in $\mathcal{L}^2$ so that $\mathsf{E}(M_n^2) < \infty$, $\forall n$. Then for $s, t, u, v \in \mathbf{Z}^+$, with $s \leq t \leq u \leq v$, we know that

$$\mathsf{E}(M_v | \mathcal{F}_u) = M_u \quad \text{(a.s.)},$$

so that $M_v - M_u$ is orthogonal to $\mathcal{L}^2(\mathcal{F}_u)$ (see Section 9.5) and in particular,

(a) $$\langle M_t - M_s, M_v - M_u \rangle = 0.$$

Hence the formula

$$M_n = M_0 + \sum_{k=1}^{n}(M_k - M_{k-1})$$

expresses M_n as the sum of orthogonal terms, and Pythagoras's theorem yields

(b) $$\mathsf{E}(M_n^2) = \mathsf{E}(M_0^2) + \sum_{k=1}^{n} \mathsf{E}[(M_k - M_{k-1})^2].$$

THEOREM

▶   *Let M be a martingale for which $M_n \in \mathcal{L}^2$, $\forall n$. Then M is bounded in $\mathcal{L}^2$ if and only if*

(c) $$\sum \mathsf{E}[(M_k - M_{k-1})^2] < \infty;$$

and when this obtains,

$$M_n \to M_\infty \text{ almost surely and in } \mathcal{L}^2.$$

Proof. It is obvious from (b) that condition (c) is equivalent to the statement M is bounded in $\mathcal{L}^2$.

Suppose now that (c) holds. Then M is bounded in $\mathcal{L}^2$, and hence, by the property of monotonicity of norms (Section 6.7), M is bounded in $\mathcal{L}^1$. Doob's Convergence Theorem 11.5 shows that $M_\infty := \lim M_n$ exists almost surely. The Pythagorean theorem implies that

(d) $$\mathsf{E}[(M_{n+r} - M_n)^2] = \sum_{k=n+1}^{n+r} \mathsf{E}[(M_k - M_{k-1})^2].$$

Letting $r \to \infty$ and applying Fatou's Lemma, we obtain

(e) $$\mathsf{E}[(M_\infty - M_n)^2] \le \sum_{k \ge n+1} \mathsf{E}[(M_k - M_{k-1})^2].$$

Hence

(f) $$\lim_{n} \mathsf{E}[(M_\infty - M_n)^2] = 0,$$

so that $M_n \to M_\infty$ in $\mathcal{L}^2$. Of course, (f) allows us to deduce from (d) that (e) *holds with equality.*

12.2. Sums of zero-mean independent variables in $\mathcal{L}^2$

THEOREM

▶▶   *Suppose that $(X_k : k \in \mathbb{N})$ is a sequence of independent random variables such that, for every k,*

$$\mathsf{E}(X_k) = 0, \quad \sigma_k^2 := \mathrm{Var}(X_k) < \infty.$$

(a)   *Then*

$$\left(\sum \sigma_k^2 < \infty\right) \quad \text{implies that} \quad \left(\sum X_k \text{ converges, a.s.}\right).$$

(b)   *If the variables (X_k) are bounded by some constant K in $[0, \infty)$ in that $|X_k(\omega)| \leq K$, $\forall k$, $\forall \omega$, then*

$$\left(\sum X_k \text{ converges, a.s.}\right) \quad \text{implies that} \quad \left(\sum \sigma_k^2 < \infty\right).$$

Note. Of course, the Kolmogorov 0-1 law implies that

$$\mathsf{P}(\textstyle\sum X_k \text{ converges}) = 0 \text{ or } 1.$$

Notation. We define

$$\mathcal{F}_n := \sigma(X_1, X_2, \ldots, X_n), \quad M_n := X_1 + X_2 + \cdots + X_n,$$

(with $\mathcal{F}_0 := \{\emptyset, \Omega\}$, $M_0 := 0$, by the usual conventions). We also define

$$A_n := \sum_{k=1}^{n} \sigma_k^2, \quad N_n := M_n^2 - A_n,$$

so that $A_0 := 0$ and $N_0 := 0$.

Proof of (a). We know from (10.4,a) that M is a martingale. Moreover

$$(*) \qquad \mathsf{E}[(M_k - M_{k-1})^2] = \mathsf{E}(X_k^2) = \sigma_k^2,$$

so that, from (12.1,b),

$$\mathsf{E}(M_n^2) = \sum_{k=1}^{n} \sigma_k^2 = A_n.$$

If $\sum \sigma_k^2 < \infty$, then M is bounded in $\mathcal{L}^2$, so that $\lim M_n$ exists a.s.   □

Proof of (b). We can strengthen $(*)$ as follows: since X_k is independent of $\mathcal{F}_{k-1}$, we have, almost surely,

$$\mathsf{E}[(M_k - M_{k-1})^2|\mathcal{F}_{k-1}] = \mathsf{E}[X_k^2|\mathcal{F}_{k-1}] = \mathsf{E}(X_k^2) = \sigma_k^2.$$

A familiar argument now applies: since M_{k-1} is $\mathcal{F}_{k-1}$ measurable,

$$\begin{aligned}
\sigma_k^2 &= \mathsf{E}(M_k^2|\mathcal{F}_{k-1}) - 2M_{k-1}\mathsf{E}(M_k|\mathcal{F}_{k-1}) + M_{k-1}^2 \\
&= \mathsf{E}(M_k^2|\mathcal{F}_{k-1}) - M_{k-1}^2 \quad \text{(a.s.)}
\end{aligned}$$

But this result states that

$$N \text{ is a martingale.}$$

Now let $c \in (0, \infty)$ and define

$$T := \inf\{r : |M_r| > c\}.$$

We know that N^T is a martingale so that, for every n,

$$\mathsf{E}N_n^T = \mathsf{E}[(M_n^T)^2] - \mathsf{E}A_{T \wedge n} = 0.$$

But since $|M_T - M_{T-1}| = |X_T| \leq K$ if T is finite, we see that $|M_n^T| \leq K + c$ for every n, whence

$(**)$ $$\mathsf{E}A_{T \wedge n} \leq (K + c)^2, \quad \forall n.$$

However, since $\sum X_n$ converges a.s., the partial sums of $\sum X_k$ are a.s. bounded, and it must be the case that for some c, $\mathsf{P}(T = \infty) > 0$. It is now clear from $(**)$ that $A_\infty := \sum \sigma_k^2 < \infty$. $\qquad \square$

Remark. The proof of (b) showed that if (X_n) is a sequence of independent zero-mean RVs uniformly bounded by some constant K, then

$$(\mathsf{P}\{ \text{ partial sums of } \sum X_k \text{ are bounded } \} > 0) \Rightarrow (\sum X_k \text{ converges a.s.})$$

Generalization. Sections 12.11-12.16 present the natural martingale form of Theorem 12.2 with applications.

12.3. Random signs

Suppose that (a_n) is a sequence of real numbers and that (ε_n) is a sequence of IID RVs with

$$\mathsf{P}(\varepsilon_n = \pm 1) = \tfrac{1}{2}.$$

The results of Section 12.2 show that

$$\sum \varepsilon_n a_n \text{ converges (a.s.) } \textit{if and only if } \sum a_n^2 < \infty,$$

and that

$$\sum \varepsilon_n a_n \text{ (a.s.) } \textit{oscillates infinitely if } \sum a_n^2 = \infty.$$

You should think about how to clinch the latter statement.

12.4. A symmetrization technique: expanding the sample space

We need a stronger result than that provided by (12.2,b).

LEMMA

Suppose that (X_n) is a sequence of independent random variables bounded by a constant K in $[0, \infty)$:

$$|X_n(\omega)| \le K, \quad \forall n, \forall \omega.$$

Then

$$(\sum X_n \text{ converges, a.s.}) \Rightarrow (\sum \mathsf{E}(X_n) \text{ converges and } \sum \text{Var}(X_n) < \infty).$$

Proof. If each X_n has mean zero, then of course, this would amount to (12.2,b). There is a nice trick which replaces each X_n by a 'symmetrized version' Z_n^* of mean 0 in such a way as to preserve enough of the structure. Let $(\tilde{\Omega}, \tilde{\mathcal{F}}, \tilde{\mathsf{P}}, (\tilde{X}_n : n \in \mathbb{N}))$ be an exact copy of $(\Omega, \mathcal{F}, \mathsf{P}, (X_n : n \in \mathbb{N}))$. Define

$$(\Omega^*, \mathcal{F}^*, \mathsf{P}^*) := (\Omega, \mathcal{F}, \mathsf{P}) \times (\tilde{\Omega}, \tilde{\mathcal{F}}, \tilde{\mathsf{P}})$$

and, for $\omega^* = (\omega, \tilde{\omega}) \in \Omega^*$, define

$$X_n^*(\omega^*) := X_n(\omega), \quad \tilde{X}_n^*(\omega^*) := \tilde{X}_n(\tilde{\omega}), \quad Z_n^*(\omega^*) := X_n^*(\omega^*) - \tilde{X}_n(\omega^*).$$

We think of X_n^* as X_n lifted to the larger 'sample space' $(\Omega^*, \mathcal{F}^*, \mathsf{P}^*)$. It is clear (and may be proved by applying Uniqueness Lemma 1.6 in a familiar way) that the combined family

$$(X_n^* : n \in \mathbb{N}) \cup (\tilde{X}_n^* : n \in \mathbb{N})$$

is a family of independent random variables on $(\Omega^*, \mathcal{F}^*, \mathsf{P}^*)$, with both X_n^* and $\tilde{X}_n^*$ having the same P^*-distribution as the P-distribution of X_n:

$$\mathsf{P}^* \circ (X_n^*)^{-1} = \mathsf{P} \circ X_n^{-1} \text{ on } (\mathbf{R}, \mathcal{B}), \text{ etc.}$$

Now we have

(a) $(Z_n^* : n \in \mathbf{N}^*)$ *is a* **zero-mean** *sequence of independent random variables on* $(\Omega^*, \mathcal{F}^*, \mathbf{P}^*)$ *such that* $|Z_n(\omega^*)| \leq 2K$ $(\forall n, \forall \omega^*)$ *and*

$$\text{Var}(Z_n^*) = 2\sigma_n^2,$$

where $\sigma_n^2 := \text{Var}(X_n)$.

  Let
$$G := \{\omega \in \Omega : \sum X_n(\omega) \text{ converges}\},$$

with $\tilde{G}$ defined similarly. Then we are given that $\mathbf{P}(G) = \tilde{\mathbf{P}}(\tilde{G}) = 1$, so that $\mathbf{P}^*(G \times \tilde{G}) = 1$. But $\sum Z_n^*(\omega^*)$ converges on $G \times \tilde{G}$, so that

(b) $\mathbf{P}^*(\sum Z_n^* \text{ converges}) = 1.$

From (a) and (b) and (12.2,b), we conclude that

$$\sum \sigma_n^2 < \infty,$$

and now it follows from (12.2,a) that

(c) $\sum [X_n - \mathbf{E}(X_n)] \text{ converges, a.s.},$

the variables in this sum being zero-mean independent, with

$$\mathbf{E}[\{X_n - \mathbf{E}(X_n)\}^2] = \sigma_n^2.$$

Since (c) holds and $\sum X_n$ converges (a.s.) by hypothesis, $\sum \mathbf{E}(X_n)$ converges. $\square$

Note. Another proof of the lemma may be found in Section 18.6.

12.5. Kolmogorov's Three-Series Theorem

Let (X_n) *be a sequence of independent random variables. Then* $\sum X_n$ *converges almost surely* **if and only if** *for some (then for every)* $K > 0$, *the following three properties hold:*

  (i) $\sum_n \mathbf{P}(|X_n| > K) < \infty,$

  (ii) $\sum_n \mathbf{E}(X_n^K)$ *converges,*

  (iii) $\sum_n \text{Var}(X_n^K) < \infty,$

where
$$X_n^K(\omega) := \begin{cases} X_n(\omega) & \text{if } |X_n(\omega)| \le K, \\ 0 & \text{if } |X_n(\omega)| > K. \end{cases}$$

Proof of 'if' part. Suppose that for *some* $K > 0$ properties (i)-(iii) hold. Then
$$\sum \mathsf{P}(X_n \neq X_n^K) = \sum \mathsf{P}(|X_n| > K) < \infty,$$
so that by (BC1)
$$\mathsf{P}(X_n = X_n^K \text{ for all but finitely many } n) = 1.$$

It is therefore clear that we need only show that $\sum X_n^K$ converges almost surely; and because of (ii), we need only prove that
$$\sum Y_n^K \text{ converges, a.s., where } Y_n^K := X_n^K - \mathsf{E}(X_n^K).$$

However, the sequence $(Y_n^K : n \in \mathsf{N})$ is a zero-mean sequence of independent random variables with
$$\mathsf{E}[(Y_n^K)^2] = \operatorname{Var}(X_n^K).$$

Because of (iii), the desired result now follows from (12.2,a). $\square$

Proof of 'only if' part. Suppose that $\sum X_n$ converges, a.s., and that K is *any* constant in $(0, \infty)$. Since it is almost surely true that $X_n \to 0$ whence $|X_n| > K$ for only finitely many n, (BC2) shows that (i) holds. Since (a.s.) $X_n = X_n^K$ for all but finitely many n, we know that
$$\sum X_n^K \text{ converges, a.s.}$$

Lemma 12.4 completes the proof. $\square$

Results such as the Three-Series Theorem become powerful when used in conjunction with Kronecker's Lemma (Section 12.7).

12.6. Cesàro's Lemma

Suppose that (b_n) is a sequence of strictly positive real numbers with $b_n \uparrow \infty$, and that (v_n) is a convergent sequence of real numbers: $v_n \to v_\infty \in \mathsf{R}$. Then
$$\frac{1}{b_n} \sum_{k=1}^{n} (b_k - b_{k-1}) v_k \to v_\infty \quad (n \to \infty).$$

Here, $b_0 := 0$.

Proof. Let $\varepsilon > 0$. Choose N such that

$$v_k > v_\infty - \varepsilon \text{ whenever } k \geq N.$$

Then,

$$\liminf_{n \to \infty} \frac{1}{b_n} \sum_{k=1}^{n} (b_k - b_{k-1}) v_k$$

$$\geq \liminf \left\{ \frac{1}{b_n} \sum_{k=1}^{N} (b_k - b_{k-1}) v_k + \frac{b_n - b_N}{b_n} (v_\infty - \varepsilon) \right\}$$

$$\geq 0 + v_\infty - \varepsilon.$$

Since this is true for every $\varepsilon > 0$, we have $\liminf \geq v_\infty$; and since, by a similar argument, $\limsup \leq v_\infty$, the result follows. $\square$

12.7. Kronecker's Lemma

▶ *Again, let (b_n) denote a sequence of strictly positive real numbers with $b_n \uparrow \infty$. Let (x_n) be a sequence of real numbers, and define*

$$s_n := x_1 + x_2 + \cdots + x_n.$$

Then

$$\left(\sum \frac{x_n}{b_n} \text{ converges} \right) \Rightarrow \left(\frac{s_n}{b_n} \to 0 \right).$$

Proof. Let $u_n := \sum_{k \leq n} (x_k / b_k)$, so that $u_\infty := \lim u_n$ exists. Then

$$u_n - u_{n-1} = x_n / b_n.$$

Thus

$$s_n = \sum_{k=1}^{n} b_k (u_k - u_{k-1}) = b_n u_n - \sum_{k=1}^{n} (b_k - b_{k-1}) u_{k-1}.$$

Cesàro's Lemma now shows that

$$s_n / b_n \to u_\infty - u_\infty = 0. \qquad \square$$

12.8. A Strong Law under variance constraints

LEMMA

Let (W_n) be a sequence of independent random variables such that

$$E(W_n) = 0, \quad \sum \frac{\text{Var}(W_n)}{n^2} < \infty.$$

Then $n^{-1}\sum_{k \leq n} W_k \to 0$, a.s..

Proof. By Kronecker's Lemma, it is enough to prove that $\sum(W_n/n)$ converges, a.s. But this is immediate from Theorem 12.2(a). $\qquad\square$

Note. We are now going to see that a truncation technique enables us to obtain the general Strong Law for IID RVs from the above lemma.

12.9. Kolmogorov's Truncation Lemma

Suppose that $X_1, X_2, \ldots$ are IID RVs each with the same distribution as X, where $E(|X|) < \infty$. Set $\mu := E(X)$. Define

$$Y_n := \begin{cases} X_n & \text{if } |X_n| \leq n, \\ 0 & \text{if } |X_n| > n. \end{cases}$$

Then

  (i) $E(Y_n) \to \mu$;

  (ii) $P[Y_n = X_n \text{ eventually}] = 1$;

  (iii) $\sum n^{-2}\text{Var}(Y_n) < \infty.$

Proof of (i). Let

$$Z_n := \begin{cases} X & \text{if } |X| \leq n, \\ 0 & \text{if } |X| > n. \end{cases}$$

Then Z_n has the same distribution as Y_n, so that in particular, $E(Z_n) = E(Y_n)$. But, as $n \to \infty$, we have

$$Z_n \to X, \quad |Z_n| \leq |X|,$$

so, by (DOM), $E(Z_n) \to E(X) = \mu$. $\qquad\square$

Proof of (ii). We have

$$\sum_{n=1}^{\infty} P(Y_n \neq X_n) = \sum P(|X_n| > n) = \sum P(|X| > n)$$

$$= E\sum_{n=1}^{\infty} I_{\{|X|>n\}} = E\sum_{1\le n<|X|} 1 \le E(|X|) < \infty,$$

so that by (BC1), result (ii) holds. □

Proof of (iii). We have

$$\sum \frac{\mathrm{Var}(Y_n)}{n^2} \le \sum \frac{E(Y_n^2)}{n^2} = \sum_n \frac{E(|X|^2; |X| \le n)}{n^2} = E[|X|^2 f(|X|)],$$

where, for $0 < z < \infty$,

$$f(z) = \sum_{n\ge\max(1,z)} n^{-2} \le 2/\max(1,z).$$

We have used the fact that, for $n \ge 1$,

$$\frac{1}{n^2} \le \frac{2}{n(n+1)} = 2\left(\frac{1}{n} - \frac{1}{n+1}\right).$$

Hence

$$\sum n^{-2}\mathrm{Var}(Y_n) \le 2E(|X|) < \infty. \qquad \square$$

12.10. Kolmogorov's Strong Law of Large Numbers (SLLN)

▶▶   *Let $X_1, X_2, \ldots$ be IID RVs with $E(|X_k|) < \infty$, $\forall k$. Define*

$$S_n := X_1 + X_2 + \cdots + X_n.$$

Then, with $\mu := E(X_k)$, $\forall k$,

$$n^{-1}S_n \to \mu, \quad \text{almost surely.}$$

Proof. Define Y_n as in Lemma 12.9. By property (ii) of that lemma, we need only show that

$$n^{-1}\sum_{k\le n} Y_k \to \mu, \quad \text{a.s.}$$

But

(a)
$$n^{-1} \sum_{k \leq n} Y_k = n^{-1} \sum_{k \leq n} \mathsf{E}(Y_k) + n^{-1} \sum_{k \leq n} W_k,$$

where $W_k := Y_k - \mathsf{E}(Y_k)$. But, the first term on the right-hand side of (a) tends to μ by (12.9,i) and Cesàro's Lemma; and the second almost surely converges to 0 by Lemma 12.8. $\qquad\square$

Notes. The Strong Law is philosophically satisfying in that it gives a precise formulation of $\mathsf{E}(X)$ as 'the mean of a large number of independent realizations of X'. We know from Exercise E4.6 that if $\mathsf{E}(|X|) = \infty$, then

$$\limsup |S_n|/n = \infty, \quad \text{a.s.}.$$

So, we have arrived at the best possible result for the IID case.

Discussion of methods. Even though we have achieved a good result, it has to be admitted that the truncation technique seems '*ad hoc*': it does not have the pure-mathematical elegance – the sense of rightness – which the martingale proof and the proof by ergodic theory (the latter is not in this book) both possess. However, *each of the methods can be adapted to cover situations which the others cannot tackle; and, in particular, classical truncation arguments retain great importance.*

Properly formulated, the argument which gave the result, Theorem 12.2, on which all of this chapter has so far relied, can yield much more.

12.11. Doob decomposition

In the following theorem, the statement that 'A is a previsible process null at 0' means of course that $A_0 = 0$ and $A_n \in m\mathcal{F}_{n-1}$ $(n \in \mathbb{N})$.

THEOREM

▶▶(a) Let $(X_n : n \in \mathbb{Z}^+)$ be an adapted process with $X_n \in \mathcal{L}^1, \forall n$. Then X has a Doob decomposition

(D)
$$X = X_0 + M + A$$

where M is a martingale null at 0, and A is a previsible process null at 0. Moreover, this decomposition is unique modulo indistinguishability in the sense that if $X = X_0 + \tilde{M} + \tilde{A}$ is another such decomposition, then

$$\mathsf{P}(M_n = \tilde{M}_n, A_n = \tilde{A}_n, \forall n) = 1.$$

▶(b)   *X is a submartingale if and only if the process A is an increasing process in the sense that*

$$\mathsf{P}(A_n \leq A_{n+1}, \forall n) = 1.$$

Proof. If X has a Doob decomposition as at (D), then, since M is a martingale and A is previsible, we have, almost surely,

$$\mathsf{E}(X_n - X_{n-1}|\mathcal{F}_{n-1}) = \mathsf{E}(M_n - M_{n-1}|\mathcal{F}_{n-1}) + \mathsf{E}(A_n - A_{n-1}|\mathcal{F}_{n-1})$$
$$= 0 + (A_n - A_{n-1}).$$

Hence

(c) $$A_n = \sum_{k=1}^{n} \mathsf{E}(X_k - X_{k-1}|\mathcal{F}_{k-1}), \quad \text{a.s.,}$$

and if we use (c) to *define* A, we obtain the required decomposition of X. The 'submartingale' result (b) is now obvious.   □

Remark. The **Doob-Meyer decomposition**, which expresses a submartingale in *continuous* time as the sum of a local martingale and a previsible increasing process, is a deep result which is the foundation stone for **stochastic-integral** theory.

12.12. The angle-brackets process $\langle M \rangle$

Let M be a martingale in $\mathcal{L}^2$ and null at 0. Then the conditional form of Jensen's inequality shows that

(a) M^2 *is a submartingale.*

Thus M has a Doob decomposition (essentially unique):

(b) $M^2 = N + A,$

where N is a martingale and A is a previsible increasing process, both N and A being null at 0. Define $A_\infty := \uparrow \lim A_n$, a.s.

Notation. *The process A is often written $\langle M \rangle$.*

Since $E(M_n^2) = E(A_n)$, we see that

(c)   *M is bounded in $\mathcal{L}^2$ if and only if $E(A_\infty) < \infty$.*

It is important to note that

▶(d) $A_n - A_{n-1} = E(M_n^2 - M_{n-1}^2 | \mathcal{F}_{n-1}) = E[(M_n - M_{n-1})^2 | \mathcal{F}_{n-1}].$

12.13. Relating convergence of M to finiteness of $\langle M \rangle_\infty$

Again let M be a martingale in $\mathcal{L}^2$ and null at 0. Define $A := \langle M \rangle$. (More strictly, let A be 'a version of' $\langle M \rangle$.)

THEOREM

▶(a) $\lim_n M_n(\omega)$ *exists for almost every ω for which $A_\infty(\omega) < \infty$.*

▶(b)   *Suppose that M has uniformly bounded increments in that for some K in $\mathbf{R}$,*

$$|M_n(\omega) - M_{n-1}(\omega)| \leq K, \quad \forall n, \forall \omega.$$

Then $A_\infty(\omega) < \infty$ for almost every ω for which $\lim_n M_n(\omega)$ exists.

Remark. This is obviously an extension – and a very substantial one – of Theorem 12.2.

Proof of (a). Because A is *previsible*, it is immediate that for every $k \in \mathbf{N}$,

$$S(k) := \inf\{n \in \mathbf{Z}^+ : A_{n+1} > k\}$$

defines a stopping time $S(k)$. Moreover, *the stopped process $A^{S(k)}$ is previsible* because for $B \in \mathcal{B}$, and $n \in \mathbf{N}$,

$$\{A_{n \wedge S(k)} \in B\} = F_1 \cup F_2,$$

where

$$F_1 := \bigcup_{r=0}^{n-1} \{S(k) = r; A_r \in B\} \in \mathcal{F}_{n-1},$$
$$F_2 := \{A_n \in B\} \cap \{S(k) \leq n-1\}^c \in \mathcal{F}_{n-1}.$$

Since

$$(M^{S(k)})^2 - A^{S(k)} = (M^2 - A)^{S(k)}$$

is a martingale, we now see that $\langle M^{S(k)} \rangle = A^{S(k)}$. However, the process $A^{S(k)}$ is bounded by k, so that by (12.12,c), $M^{S(k)}$ is bounded in $\mathcal{L}^2$ and

(c) $\lim_n M_{n \wedge S(k)}$ *exists almost surely.*

However,

(d) $$\{A_\infty < \infty\} = \bigcup_k \{S(k) = \infty\}.$$

Result (a) now follows on combining (c) and (d). □

Proof of (b). Suppose that

$$P(A_\infty = \infty, \sup_n |M_n| < \infty) > 0.$$

Then for some $c > 0$,

(e) $$\mathbf{P}(T(c) = \infty, A_\infty = \infty) > 0,$$

where $T(c)$ is the stopping time:

$$T(c) := \inf\{r : |M_r| > c\}.$$

Now,

$$\mathbf{E}(M_{T(c)\wedge n}^2 - A_{T(c)\wedge n}) = 0,$$

and $M^{T(c)}$ is bounded by $c + K$. Thus

(f) $$\mathbf{E}A_{T(c)\wedge n} \le (c + K)^2, \quad \forall n.$$

But (MON) shows that (e) and (f) are incompatible. Result (b) follows. □

Remark. In the proof of (a), we were able to use previsibility to make the jump $A_{S(k)} - A_{S(k)-1}$ irrelevant. We could not do this for the jump $M_{T(c)} - M_{T(c)-1}$ which is why we needed the assumption about bounded increments.

12.14. A trivial 'Strong Law' for martingales in $\mathcal{L}^2$

Let M be a martingale in $\mathcal{L}^2$ and null at 0, and let $A = \langle M \rangle$. Since $(1+A)^{-1}$ is a bounded previsible process,

$$W_n := \sum_{1 \le k \le n} \frac{M_k - M_{k-1}}{1 + A_k} = ((1+A)^{-1} \bullet M)_n$$

defines a martingale W. Moreover, since $(1 + A_n)$ is $\mathcal{F}_{n-1}$ measurable,

$$\mathbf{E}[(W_n - W_{n-1})^2 | \mathcal{F}_{n-1}] = (1 + A_n)^{-2}(A_n - A_{n-1})$$
$$\le (1 + A_{n-1})^{-1} - (1 + A_n)^{-1}, \quad \text{a.s.}$$

We see that $\langle W \rangle_\infty \leq 1$, a.s., so that $\lim W_n$ exists, a.s.. Kronecker's Lemma now shows that

▶▶(a) $M_n/A_n \to 0$ a.s. on $\{A_\infty = \infty\}$.

12.15. Lévy's extension of the Borel-Cantelli Lemmas

THEOREM

Suppose that for $n \in \mathsf{N}$, $E_n \in \mathcal{F}_n$. Define

$$Z_n := \sum_{1 \leq k \leq n} I_{E_k} = \text{number of } E_k (k \leq n) \text{ which occur.}$$

Define $\xi_k := \mathsf{P}(E_k | \mathcal{F}_{k-1})$, and

$$Y_n := \sum_{1 \leq k \leq n} \xi_k.$$

Then, almost surely,

(a) $(Y_\infty < \infty) \Rightarrow (Z_\infty < \infty)$,

(b) $(Y_\infty = \infty) \Rightarrow (Z_n/Y_n \to 1)$.

Remarks. (i) Since $\mathsf{E}\xi_k = \mathsf{P}(E_k)$, it follows that if $\sum \mathsf{P}(E_k) < \infty$, then $Y_\infty < \infty$, a.s. (BC1) therefore follows.

(ii) Let $(E_n : n \in \mathsf{N})$ be a sequence of independent events associated with some triple $(\Omega, \mathcal{F}, \mathsf{P})$, and define $\mathcal{F}_n = \sigma(E_1, E_2, \ldots, E_n)$. Then $\xi_k = \mathsf{P}(E_k)$, a.s., and (BC2) follows from (b).

Proof. Let M be the martingale $Z - Y$, so that $Z = M + Y$ is the Doob decomposition of the submartingale Z. Then (you check!)

$$A_n := \langle M \rangle_n = \sum_{k \leq n} \xi_k (1 - \xi_k) \leq Y_n, \quad \text{a.s.}$$

If $Y_\infty < \infty$, then $A_\infty < \infty$ and $\lim M_n$ exists, so that Z_∞ is finite. (We are skipping 'except for a null ω-set' statements now.)

If $Y_\infty = \infty$ and $A_\infty < \infty$ then $\lim M_n$ exists and it is trivial that $Z_n/Y_n \to 1$.

If $Y_\infty = \infty$ and $A_\infty = \infty$, then $M_n/A_n \to 0$, so that, *a fortiori*, $M_n/Y_n \to 0$ and $Z_n/Y_n \to 1$.   □

12.16. Comments

The last few sections have indicated just how powerful the use of $\langle M \rangle$ to study M is likely to be. In the same way as one can obtain the conditional version Theorem 12.15 of the Borel-Cantelli Lemmas, one can obtain conditional versions of the Three-Series Theorem etc. But a whole *new* world is opened up: see Neveu (1975), for example. In the continuous-time case, things are much more striking still. See, for example, Rogers and Williams (1987).

Chapter 13
Uniform Integrability

We have already seen a number of nice applications of martingale theory. To derive the full benefit, we need something better than the Dominated-Convergence Theorem. In particular, Theorem 13.7 gives a necessary and sufficient condition for a sequence of RVs to converge on $\mathcal{L}^1$. The new concept required is that of a **uniformly integrable (UI) family of random variables.** *This concept links perfectly with conditional expectations and hence with martingales.*

The appendix to this chapter contains a discussion of that topic loved by examiners and others: **modes of convergence.** Our use of the Upcrossing Lemma has meant that this topic does not feature large in the main text of this book.

13.1. An 'absolute continuity' property

LEMMA

▶(a) *Suppose that $X \in \mathcal{L}^1 = \mathcal{L}^1(\Omega, \mathcal{F}, \mathbf{P})$. Then, given $\varepsilon > 0$, there exists a $\delta > 0$ such that for $F \in \mathcal{F}$, $\mathbf{P}(F) < \delta$ implies that $\mathbf{E}(|X|; F) < \varepsilon$.*

Proof. If the conclusion is false, then, for some $\varepsilon_0 > 0$, we can find a sequence (F_n) of elements of $\mathcal{F}$ such that

$$\mathbf{P}(F_n) < 2^{-n} \quad \text{and} \quad \mathbf{E}(|X|; F_n) \geq \varepsilon_0.$$

Let $H := \limsup F_n$. Then (BC1) shows that $\mathbf{P}(H) = 0$, but the 'Reverse' Fatou Lemma (5.4,b) shows that

$$\mathbf{E}(|X|; H) \geq \varepsilon_0;$$

and we have arrived at the required contradiction. $\qquad\square$

Corollary

(b)   *Suppose that $X \in \mathcal{L}^1$ and that $\varepsilon > 0$. Then there exists K in $[0, \infty)$ such that*

$$E(|X|; |X| > K) < \varepsilon.$$

Proof. Let δ be as in Lemma (a). Since

$$K P(|X| > K) \le E(|X|),$$

we can choose K such that $P(|X| > K) < \delta$. $\square$

13.2. Definition. UI family

▶▶   *A class $\mathcal{C}$ of random variables is called* **uniformly integrable (UI)** *if given $\varepsilon > 0$, there exists K in $[0, \infty)$ such that*

$$E(|X|; |X| > K) < \varepsilon, \quad \forall X \in \mathcal{C}.$$

We note that for such a class $\mathcal{C}$, we have (with K_1 relating to $\varepsilon = 1$) for every $X \in \mathcal{C}$,

$$\begin{aligned} E(|X|) &= E(|X|; |X| > K_1) + E(|X|; |X| \le K_1) \\ &\le 1 + K_1, \end{aligned}$$

Thus, *a UI family is bounded in $\mathcal{L}^1$.*

It is not true that a family bounded in $\mathcal{L}^1$ is UI.

Example. Take $(\Omega, \mathcal{F}, P) = ([0,1], \mathcal{B}[0,1], \text{Leb})$. Let

$$E_n = (0, n^{-1}), \quad X_n = n I_{E_n}.$$

Then $E(|X_n|) = 1$, $\forall n$, so that (X_n) is bounded in $\mathcal{L}^1$. However, for any $K > 0$, we have for $n > K$,

$$E(|X_n|; |X_n| > K) = n P(E_n) = 1,$$

so that (X_n) is not UI. Here, $X_n \to 0$, but $E(X_n) \not\to 0$.

13.3. Two simple sufficient conditions for the UI property

▶(a)   *Suppose that $\mathcal{C}$ is a class of random variables which is bounded in $\mathcal{L}^p$ for some $p > 1$; thus, for some $A \in [0, \infty)$,*

$$E(|X|^p) < A, \quad \forall X \in \mathcal{C}.$$

Then C is UI.

Proof. If $v \geq K > 0$, then $v \leq K^{1-p}v^p$ (obviously!). Hence, for $K > 0$ and $X \in C$, we have

$$\mathsf{E}(|X|; |X| > K) \leq K^{1-p}\mathsf{E}(|X|^p; |X| > K) \leq K^{1-p}A.$$

The result follows.   □

(b)   *Suppose that C is a class of random variables which is dominated by an integrable non-negative variable Y:*

$$|X(\omega)| \leq Y(\omega), \quad \forall X \in C \quad and \quad \mathsf{E}(Y) < \infty.$$

Then C is UI.

Note. It is precisely this which makes (DOM) work for our $(\Omega, \mathcal{F}, \mathsf{P})$.

Proof. It is obvious that, for $K > 0$ and $X \in C$,

$$\mathsf{E}(|X|; |X| > K) \leq \mathsf{E}(Y; Y > K),$$

and now it is only necessary to apply (13.1,b) to Y.   □

13.4. UI property of conditional expectations

The mean reason that the UI property fits in so well with martingale theory is the following. See Exercise E13.3 for an important extension.

THEOREM

▶▶   *Let $X \in \mathcal{L}^1$. Then the class*

$$\{\mathsf{E}(X|\mathcal{G}) : \mathcal{G} \text{ a sub-}\sigma\text{-algebra of } \mathcal{F}\}$$

is uniformly integrable.

Note. Because of the business of versions, a formal description of the class C in question would be as follows: $Y \in C$ if and only if for some sub-σ-algebra $\mathcal{G}$ of $\mathcal{F}$, Y is a version of $\mathsf{E}(X|\mathcal{G})$.

Proof. Let $\varepsilon > 0$ be given. Choose $\delta > 0$ such that, for $F \in \mathcal{F}$,

$$\mathsf{P}(F) < \delta \text{ implies that } \mathsf{E}(|X|; F) < \varepsilon.$$

Choose K so that $K^{-1}\mathsf{E}(|X|) < \delta$.

Now let $\mathcal{G}$ be a sub-σ-algebra of $\mathcal{F}$ and let Y be any version of $\mathsf{E}(X|\mathcal{G})$. By Jensen's inequality,

(a) $$|Y| \leq \mathsf{E}(|X| \,|\, \mathcal{G}), \quad \text{a.s.}$$

Hence $\mathsf{E}(|Y|) \leq \mathsf{E}(|X|)$ and

$$K\mathsf{P}(|Y| > K) \leq \mathsf{E}(|Y|) \leq \mathsf{E}(|X|),$$

so that

$$\mathsf{P}(|Y| > K) < \delta.$$

But $\{|Y| > K\} \in \mathcal{G}$, so that, from (a) and the definition of conditional expectation,

$$\mathsf{E}(|Y|; |Y| \geq K) \leq \mathsf{E}(|X|; |Y| \geq K) < \varepsilon. \qquad \square$$

Note. Now you can see why we needed the more subtle result (13.1,a), not just the result (13.1,b) which has a simpler proof.

13.5. Convergence in probability

Let (X_n) be a sequence of random variables, and let X be a random variable. We say that

▶▶ $\qquad X_n \to X$ **in probability**

if for every $\varepsilon > 0$,

$$\mathsf{P}(|X_n - X| > \varepsilon) \to 0 \text{ as } n \to \infty.$$

LEMMA

▶ $\qquad$ *If $X_n \to X$ almost surely, then*

$$X_n \to X \text{ in probability.}$$

Proof. Suppose that $X_n \to X$ almost surely and that $\varepsilon > 0$. Then by the Reverse Fatou Lemma 2.6(b) for sets,

$$0 = \mathsf{P}(|X_n - X| > \varepsilon, \text{ i.o.}) = \mathsf{P}(\limsup\{|X_n - X| > \varepsilon\})$$
$$\geq \limsup \mathsf{P}(|X_n - X| > \varepsilon),$$

and the result is proved. $\qquad \square$

Note. As already mentioned, a discussion of the relationships between various modes of convergence may be found in the appendix to this chapter.

13.6. Elementary proof of (BDD)

We restate the Bounded Convergence Theorem, but under the weaker hypothesis of 'convergence in probability' rather than 'almost sure convergence'.

THEOREM (BDD)

> Let (X_n) be a sequence of RVs, and let X be a RV. Suppose that $X_n \to X$ in probability and that for some K in $[0, \infty)$, we have for every n and ω
>
> $$|X_n(\omega)| \leq K.$$
>
> Then
>
> $$E(|X_n - X|) \to 0.$$

Proof. Let us check that $P(|X| \leq K) = 1$. Indeed, for $k \in \mathbb{N}$,

$$P(|X| > K + k^{-1}) \leq P(|X - X_n| > k^{-1}), \quad \forall n,$$

so that $P(|X| > K + k^{-1}) = 0$. Thus

$$P(|X| > K) = P(\bigcup_k \{|X| > K + k^{-1}\}) = 0.$$

Let $\varepsilon > 0$ be given. Choose n_0 such that

$$P(|X_n - X| > \tfrac{1}{3}\varepsilon) < \frac{\varepsilon}{3K} \text{ when } n \geq n_0.$$

Then, for $n \geq n_0$,

$$E(|X_n - X|) = E(|X_n - X|; |X_n - X| > \tfrac{1}{3}\varepsilon) + E(|X_n - X|; |X_n - X| \leq \tfrac{1}{3}\varepsilon)$$
$$\leq 2KP(|X_n - X| > \tfrac{1}{3}\varepsilon) + \tfrac{1}{3}\varepsilon \leq \varepsilon.$$

The proof is finished. $\qquad\qquad\qquad\qquad\qquad\qquad\qquad\qquad\qquad\qquad$ □

This proof shows (much as does that of the Weierstrass approximation theorem) that convergence in probability is a natural concept.

13.7. A necessary and sufficient condition for $\mathcal{L}^1$ convergence

THEOREM

▶▶   *Let (X_n) be a sequence in $\mathcal{L}^1$, and let $X \in \mathcal{L}^1$. Then $X_n \to X$ in $\mathcal{L}^1$, equivalently $\mathsf{E}(|X_n - X|) \to 0$, if and only if the following two conditions are satisfied:*

(i) *$X_n \to X$ in probability,*

(ii) *the sequence (X_n) is UI.*

Remarks. It is of course the 'if' part of the theorem which is useful. Since the result is 'best possible', it must improve on (DOM) for our $(\Omega, \mathcal{F}, \mathbf{P})$ triple; and, of course, result 13.3(b) makes this explicit.

Proof of 'if' part. Suppose that conditions (i) and (ii) are satisfied. For $K \in [0, \infty)$, define a function $\varphi_K : \mathbf{R} \to [-K, K]$ as follows:

$$\varphi_K(x) := \begin{cases} K & \text{if } x > K, \\ x & \text{if } |x| \leq K, \\ -K & \text{if } x < -K. \end{cases}$$

Let $\varepsilon > 0$ be given. By the UI property of the (X_n) sequence and (13.1,b), we can choose K so that

$$\mathsf{E}\{|\varphi_K(X_n) - X_n|\} < \frac{\varepsilon}{3}, \forall n; \qquad \mathsf{E}\{|\varphi_K(X) - X|\} < \frac{\varepsilon}{3}.$$

But, since $|\varphi_K(x) - \varphi_K(y)| \leq |x - y|$, we see that $\varphi_K(X_n) \to \varphi_K(X)$ in probability; and by (BDD) in the form of the preceding section, we can choose n_0 such that, for $n \geq n_0$,

$$\mathsf{E}\{|\varphi_K(X_n) - \varphi_K(X)|\} < \frac{\varepsilon}{3}.$$

The triangle inequality therefore implies that, for $n \geq n_0$,

$$\mathsf{E}(|X_n - X|) < \varepsilon,$$

and the proof is complete.   □

Proof of 'only if' part. Suppose that $X_n \to X$ in $\mathcal{L}^1$. Let $\varepsilon > 0$ be given. Choose N such that

$$n \geq N \quad \Rightarrow \quad \mathsf{E}(|X_n - X|) < \varepsilon/2.$$

By (13.1,a), we can choose $\delta > 0$ such that whenever $P(F) < \delta$, we have

$$E(|X_n|; F) < \varepsilon \qquad (1 \leq n \leq N),$$
$$E(|X|; F) < \varepsilon/2.$$

Since (X_n) is bounded in $\mathcal{L}^1$, we can choose K such that

$$K^{-1} \sup_r E(|X_r|) < \delta.$$

Then for $n \geq N$, we have $P(|X_n| > K) < \delta$ and

$$E(|X_n|; |X_n| > K)$$
$$\leq E(|X|; |X_n| > K) + E(|X - X_n|) < \varepsilon.$$

For $n \leq N$, we have $P(|X_n| > K) < \delta$ and

$$E(|X_n|; |X_n| > K) < \varepsilon.$$

Hence (X_n) is a UI family.

Since

$$\varepsilon P(|X_n - X| > \varepsilon) \leq E(|X_n - X|) = \|X_n - X\|_1,$$

it is clear that $X_n \to X$ in probability. $\qquad\square$

Chapter 14
UI Martingales

14.0. Introduction

The first part of this chapter examines what happens when uniform integrability is combined with the martingale property. In addition to new results such as **Lévy's 'Upward' and 'Downward' Theorems**, we also obtain new proofs of the **Kolmogorov 0-1 Law** and of the **Strong Law of Large Numbers**.

The second part of the chapter (beginning at Section 14.6) is concerned with **Doob's Submartingale Inequality**. This result implies in particular that for $p > 1$ (but not for $p = 1$) a martingale bounded in $\mathcal{L}^p$ is *dominated* by an element of $\mathcal{L}^p$ and hence converges both almost surely and in $\mathcal{L}^p$. The Submartingale Inequality is also used to prove **Kakutani's Theorem** on product-form martingales and, in an illustration of **exponential bounds**, to prove a very special case of the **Law of the Iterated Logarithm**.

The **Radon-Nikodým** theorem is then proved, and its relevance to **likelihood ratio** explained.

The topic of **optional sampling**, important for continuous-parameter theory and in other contexts, is covered in the appendix to this chapter.

14.1. UI martingales

Let M be a UI martingale, so that M is a martingale relative to our set-up $(\Omega, \mathcal{F}, \{\mathcal{F}_n\}, \mathbf{P})$ and $(M_n : n \in \mathbf{Z}^+)$ is a UI family.

Since M is UI, M is bounded in $\mathcal{L}^1$ (by (13.2)), and so $M_\infty := \lim M_n$ exists almost surely. By Theorem 13.7, it is also true that $M_n \to M_\infty$ in $\mathcal{L}^1$:

$$\mathbf{E}(|M_n - M_\infty|) \to 0.$$

We now prove that $M_n = \mathbf{E}(M_\infty | \mathcal{F}_n)$, a.s. For $F \in \mathcal{F}_n$, and $r \geq n$, the martingale property yields

$$(*) \qquad\qquad \mathbf{E}(M_r; F) = \mathbf{E}(M_n; F).$$

But

$$|E(M_r; F) - E(M_\infty; F)| \leq E(|M_r - M_\infty|; F)$$
$$\leq E(|M_r - M_\infty|).$$

Hence, on letting $r \to \infty$ in $(*)$, we obtain

$$E(M_\infty; F) = E(M_n; F).$$

We have proved the following result.

THEOREM

▶▶   *Let M be a UI martingale. Then*

$$M_\infty := \lim M_n \ \text{exists a.s. and in } \mathcal{L}^1.$$

Moreover, for every n,

$$M_n = E(M_\infty | \mathcal{F}_n), \quad \text{a.s..}$$

The obvious extension to UI supermartingales may be proved similarly.

14.2. Lévy's 'Upward' Theorem

▶▶   *Let $\xi \in \mathcal{L}^1(\Omega, \mathcal{F}, P)$, and define $M_n := E(\xi | \mathcal{F}_n)$, a.s. Then M is a UI martingale and*

$$M_n \to \eta := E(\xi | \mathcal{F}_\infty),$$

almost surely and in $\mathcal{L}^1$.

Proof. We know that M is a martingale because of the Tower Property. We know from Theorem 13.4 that M is UI. Hence $M_\infty := \lim M_n$ exists a.s. and in $\mathcal{L}^1$, and it remains only to prove that $M_\infty = \eta$, a.s., where $\eta := E(\xi | \mathcal{F}_\infty)$.

Without loss of generality, we may (and do) assume that $\xi \geq 0$. Now consider the measures Q_1 and Q_2 on $(\Omega, \mathcal{F}_\infty)$, where

$$Q_1(F) := E(\eta; F), \quad Q_2(F) = E(M_\infty; F), \quad F \in \mathcal{F}_\infty.$$

If $F \in \mathcal{F}_n$, then since $E(\eta | \mathcal{F}_n) = E(\xi | \mathcal{F}_n)$ by the Tower Property,

$$E(\eta; F) = E(M_n; F) = E(M_\infty; F),$$

the second equality having been proved in Section 14.1. Thus $\mathbf{Q}_1$ and $\mathbf{Q}_2$ agree on the π-system (algebra!) $\bigcup \mathcal{F}_n$, and hence they agree on $\mathcal{F}_\infty$.

Both η and M_∞ are $\mathcal{F}_\infty$ measurable; more strictly, M_∞ may be taken to be $\mathcal{F}_\infty$ measurable by defining $M_\infty := \limsup M_n$ for every ω. Thus,

$$F := \{\omega : \eta > M_\infty\} \in \mathcal{F}_\infty,$$

and since $\mathbf{Q}_1(F) = \mathbf{Q}_2(F)$,

$$\mathsf{E}(\eta - M_\infty; \eta > M_\infty) = 0.$$

Hence $\mathsf{P}(\eta > M_\infty) = 0$, and similarly $\mathsf{P}(M_\infty > \eta) = 0$. $\qquad\square$

14.3. Martingale proof of Kolmogorov's 0-1 law

Recall the result.

THEOREM

Let $X_1, X_2, \ldots$ be a sequence of independent RVs. Define

$$\mathcal{T}_n := \sigma(X_{n+1}, X_{n+2}, \ldots), \quad \mathcal{T} := \bigcap_n \mathcal{T}_n.$$

Then if $F \in \mathcal{T}$, $\mathsf{P}(F) = 0$ or 1.

Proof. Define $\mathcal{F}_n := \sigma(X_1, X_2, \ldots, X_n)$. Let $F \in \mathcal{T}$, and let $\eta := I_F$. Since $\eta \in b\mathcal{F}_\infty$, Lévy's Upward Theorem shows that

$$\eta = \mathsf{E}(\eta | \mathcal{F}_\infty) = \lim \mathsf{E}(\eta | \mathcal{F}_n), \quad \text{a.s.}$$

However, for each n, η is $\mathcal{T}_n$ measurable, and hence (see Remark below) is independent of $\mathcal{F}_n$. Hence by (9.7,k),

$$\mathsf{E}(\eta | \mathcal{F}_n) = \mathsf{E}(\eta) = \mathsf{P}(F), \quad \text{a.s.}$$

Hence $\eta = \mathsf{P}(F)$, a.s.; and since η only takes the values 0 and 1, the result follows. $\qquad\square$

Remark. Of course, we have cheated to some extent in building parts of the earlier proof into the martingale statements used in the proof just given.

14.4. Lévy's 'Downward' Theorem

▶▶   *Suppose that $(\Omega, \mathcal{F}, \mathbf{P})$ is a probability triple, and that $\{\mathcal{G}_{-n} : n \in \mathbf{N}\}$ is a collection of sub-σ-algebras of $\mathcal{F}$ such that*

$$\mathcal{G}_{-\infty} := \bigcap_k \mathcal{G}_{-k} \subseteq \cdots \subseteq \mathcal{G}_{-(n+1)} \subseteq \mathcal{G}_{-n} \subseteq \cdots \subseteq \mathcal{G}_{-1}.$$

Let $\gamma \in \mathcal{L}^1(\Omega, \mathcal{F}, \mathbf{P})$ and define

$$M_{-n} := \mathbf{E}(\gamma | \mathcal{G}_{-n}).$$

Then

$$M_{-\infty} := \lim M_{-n} \text{ exists a.s. and in } \mathcal{L}^1$$

and

(*) $$M_{-\infty} = \mathbf{E}(\gamma | \mathcal{G}_{-\infty}), \quad \text{a.s.}$$

Proof. The Upcrossing Lemma applied to the martingale

$$(M_k, \mathcal{G}_k : -N \leq k \leq -1)$$

can be used exactly as in the proof of Doob's Forward Convergence Theorem to show that $\lim M_{-n}$ exists a.s. The uniform-integrability result, Theorem 13.4, shows that $\lim M_{-n}$ exists in $\mathcal{L}^1$.

That (*) holds (if you like with $M_{-\infty} := \limsup M_{-n} \in m\mathcal{G}_{-\infty}$) follows by now-familiar reasoning: for $G \in \mathcal{G}_{-\infty} \subseteq \mathcal{G}_{-r}$,

$$\mathbf{E}(\gamma; G) = \mathbf{E}(M_{-r}; G),$$

and now let $r \uparrow \infty$.   □

14.5. Martingale proof of the Strong Law

Recall the result (but add $\mathcal{L}^1$ convergence as bonus).

THEOREM

Let $X_1, X_2, \ldots$ be IID RVs, with $\mathbf{E}(|X_k|) < \infty, \forall k$. Let μ be the common value of $\mathbf{E}(X_n)$. Write

$$S_n := X_1 + X_2 + \cdots + X_n.$$

Then $n^{-1}S_n \to \mu$, a.s. and in $\mathcal{L}^1$.

Proof. Define

$$\mathcal{G}_{-n} := \sigma(S_n, S_{n+1}, S_{n+2}, \ldots), \quad \mathcal{G}_{-\infty} := \bigcap_n \mathcal{G}_{-n}.$$

We know from Section 9.11 that

$$\mathbb{E}(X_1 | \mathcal{G}_{-n}) = n^{-1} S_n, \quad \text{a.s.}$$

Hence $L := \lim n^{-1} S_n$ exists a.s. and in $\mathcal{L}^1$. For definiteness, define $L := \limsup n^{-1} S_n$ for every ω. Then for each k,

$$L = \limsup \frac{X_{k+1} + \cdots + X_{k+n}}{n}$$

so that $L \in m\mathcal{T}_k$ where $\mathcal{T}_k = \sigma(X_{k+1}, X_{k+2}, \ldots)$. By Kolmogorov's 0-1 law, $\mathbb{P}(L = c) = 1$ for some c in $\mathbb{R}$. But

$$c = \mathbb{E}(L) = \lim \mathbb{E}(n^{-1} S_n) = \mu. \qquad \square$$

Exercise. Explain how we could have deduced $\mathcal{L}^1$ convergence at (12.10). *Hint.* Recall Scheffé's Lemma 5.10, and think about how to use it.

Remarks. See Meyer (1966) for important extensions and applications of the results given so far in this chapter. These extensions include: the **Hewitt-Savage 0-1 law**, **de Finetti's theorem** on exchangeable random variables, the **Choquet-Deny theorem** on bounded harmonic functions for random walks on groups.

14.6. Doob's Submartingale Inequality

THEOREM

▶▶(a)   *Let Z be a non-negative submartingale. Then, for $c > 0$,*

$$c\mathbb{P}\left(\sup_{k \leq n} Z_k \geq c\right) \leq \mathbb{E}\left(Z_n; \sup_{k \leq n} Z_k \geq c\right) \leq \mathbb{E}(Z_n).$$

Proof. Let $F := \{\sup_{k \leq n} Z_k \geq c\}$. Then F is a *disjoint* union

$$F = F_0 \cup F_1 \cup \ldots \cup F_n,$$

where

$$F_0 := \{Z_0 \geq c\},$$
$$F_k := \{Z_0 < c\} \cap \{Z_1 < c\} \cap \ldots \cap \{Z_{k-1} < c\} \cap \{Z_k \geq c\}.$$

Now, $F_k \in \mathcal{F}_k$, and $Z_k \geq c$ on F_k. Hence,

$$\mathbb{E}(Z_n; F_k) \geq \mathbb{E}(Z_k; F_k) \geq c\mathbb{P}(F_k).$$

Summing over k now yields the result.   □

The main reason for the usefulness of the above theorem is the following.

LEMMA

▶(b)   *If M is a martingale, c is a convex function, and $\mathbb{E}|c(M_n)| < \infty$, $\forall n$,
then*

$$c(M) \text{ is a submartingale.}$$

Proof. Apply the conditional form of Jensen's inequality in Table 9.7.   □

Kolmogorov's inequality

▶   *Let $(X_n : n \in \mathbb{N})$ be a sequence of independent zero-mean RVs in $\mathcal{L}^2$.
Define $\sigma_k^2 := \mathrm{Var}(X_k)$. Write*

$$S_n := X_1 + \cdots + X_n, \quad V_n := \mathrm{Var}(S_n) = \sum_{k=1}^{n} \sigma_k^2.$$

Then, for $c > 0$,

$$c^2 \mathbb{P}\left(\sup_{k \leq n} |S_k| \geq c\right) \leq V_n.$$

Proof. We know that if we set $\mathcal{F}_n = \sigma(X_1, X_2, \ldots, X_n)$, then $S = (S_n)$ is a martingale. Now apply the Submartingale Inequality to S^2.   □

Note. Kolmogorov's inequality was the key step in the original proofs of Kolmogorov's Three-Series Theorem and Strong Law.

14.7. Law of the Iterated Logarithm: special case

Let us see how the Submartingale Inequality may be used via so-called *exponential bounds* to prove a very special case of Kolmogorov's Law of the

Iterated Logarithm which is described in Section A4.1. (You would do well to take a quick look at this proof even though it is not needed later.)

THEOREM

Let $(X_n : n \in \mathbb{N})$ be IID RVs each with the standard normal $N(0,1)$ distribution of mean 0 and variance 1. Define

$$S_n := X_1 + X_2 + \cdots + X_n.$$

Then, almost surely,

$$\limsup \frac{S_n}{(2n \log \log n)^{\frac{1}{2}}} = 1.$$

Proof. Throughout the proof, we shall write

$$h(n) := (2n \log \log n)^{\frac{1}{2}}, \qquad n \geq 3.$$

(It will be understood that integers occurring in the proof are greater than e, when this is necessary.)

Step 1: An exponential bound. Define $\mathcal{F}_n := \sigma(X_1, X_2, \ldots, X_n)$. Then S is a martingale relative to $\{\mathcal{F}_n\}$. It is well known that for $\theta \in \mathbb{R}$, $n \in \mathbb{N}$,

$$\mathbf{E}e^{\theta S_n} = e^{\frac{1}{2}\theta^2 n} < \infty.$$

The function $x \mapsto e^{\theta x}$ is convex on $\mathbb{R}$, so that

$$e^{\theta S_n} \text{ is a submartingale}$$

and, by the Submartingale Inequality, we have, for $\theta > 0$,

►► $\mathbf{P}\left(\sup_{k \leq n} S_k \geq c\right) = \mathbf{P}\left(\sup_{k \leq n} e^{\theta S_k} \geq e^{\theta c}\right) \leq e^{-\theta c}\mathbf{E}\left(e^{\theta S_n}\right).$

This is a type of exponential bound much used in modern probability theory.

In our special case, we have

$$\mathbf{P}\left(\sup_{k \leq n} S_k \geq c\right) \leq e^{-\theta c}e^{\frac{1}{2}\theta^2 n},$$

and for $c > 0$, choosing the best θ, namely c/n, we obtain

(a)
$$\mathsf{P}\left(\sup_{k \leq n} S_k \geq c\right) \leq e^{-\frac{1}{2}c^2/n}.$$

Step 2: Obtaining an upper bound. Let K be a real number with $K > 1$. (We are interested in cases when K is close to 1.) Choose $c_n := Kh(K^{n-1})$. Then

$$\mathsf{P}\left(\sup_{k \leq K^n} S_k \geq c_n\right) \leq \exp(-c_n^2/2K^n) = (n-1)^{-K}(\log K)^{-K}.$$

The First Borel-Cantelli Lemma therefore shows that, almost surely, for all large n (all $n \geq n_0(\omega)$) we have for $K^{n-1} \leq k \leq K^n$,

$$S_k \leq \sup_{k \leq K^n} S_k \leq c_n = Kh(K^{n-1}) \leq Kh(k).$$

Hence, for $K > 1$,

$$\limsup_k h(k)^{-1} S_k \leq K, \quad \text{a.s.}$$

By taking a sequence of K-values converging down to 1, we obtain

$$\limsup_k h(k)^{-1} S_k \leq 1, \quad \text{a.s.}$$

Step 3: Obtaining a lower bound. Let N be an integer with $N > 1$. (We are interested in cases when N is very large.) Let ε be a number in $(0,1)$. (Of course, ε will be small in the cases which interest us.) Write $S(r)$ for S_r, etc., when typographically more convenient. For $n \in \mathbf{N}$, define the event

$$F_n := \{S(N^{n+1}) - S(N^n) > (1-\varepsilon)h(N^{n+1} - N^n)\}.$$

Then (see Proposition 14.8(b) below),

$$\mathsf{P}(F_n) = 1 - \Phi(y) \geq (2\pi)^{-\frac{1}{2}}(y + y^{-1})^{-1}\exp(-y^2/2),$$

where

$$y = (1-\varepsilon)\{2\log\log(N^{n+1} - N^n)\}^{\frac{1}{2}}.$$

Thus, ignoring 'logarithmic terms', $\mathsf{P}(F_n)$ is roughly $(n \log N)^{-(1-\varepsilon)^2}$ so that $\sum \mathsf{P}(F_n) = \infty$. However, *the events F_n ($n \in \mathbf{N}$) are clearly independent*, so that (BC2) shows that, almost surely, infinitely many F_n occur. Thus, for infinitely many n,

$$S(N^{n+1}) > (1 - \varepsilon)h(N^{n+1} - N^n) + S(N^n).$$

But, by Step 2, $S(N^n) > -2h(N^n)$ for all large n, so that for infinitely many n, we have

$$S(N^{n+1}) > (1 - \varepsilon)h(N^{n+1} - N^n) - 2h(N^n).$$

It now follows that

$$\limsup_k h(k)^{-1} S_k \geq \limsup_n h(N^{n+1})^{-1} S(N^{n+1})$$

$$\geq (1 - \varepsilon)(1 - N^{-1})^{\frac{1}{2}} - 2N^{-\frac{1}{2}}.$$

(You should check that 'the logarithmic terms do disappear'.) The rest is obvious.　　　　　□

14.8. A standard estimate on the normal distribution

We used part of the following result in the previous section.

Proposition

　　Suppose that X has the standard normal distribution, so that, for $x \in \mathbf{R}$,

$$P(X > x) = 1 - \Phi(x) = \int_x^\infty \varphi(y)dy$$

　　where

$$\varphi(y) := (2\pi)^{-\frac{1}{2}} \exp(-\tfrac{1}{2}y^2).$$

　　Then, for $x > 0$,

(a)　　　　　　　　　$P(X > x) \leq x^{-1}\varphi(x),$

(b)　　　　　　　　　$P(X > x) \geq (x + x^{-1})^{-1}\varphi(x).$

Proof. Let $x > 0$. Since $\varphi'(y) = -y\varphi(y)$,

$$\varphi(x) = \int_x^\infty y\varphi(y)dy \geq x \int_x^\infty \varphi(y)dy,$$

yielding (a).

Since $(y^{-1}\varphi(y))' = -(1 + y^{-2})\varphi(y)$,

$$x^{-1}\varphi(x) = \int_x^\infty (1 + y^{-2})\varphi(y)dy \leq (1 + x^{-2}) \int_x^\infty \varphi(y)dy,$$

yielding (b).   □

14.9. Remarks on exponential bounds; large-deviation theory

Obtaining exponential bounds is related to the very powerful theory of *large deviations* – see Varadhan (1984), Deuschel and Stroock (1989) – which has an ever-growing number of fields of application. See Ellis (1985).

You can study exponential bounds in the very specific context of martingales in Neveu (1975), Chow and Teicher (1978), Garsia (1973), etc.

Much of the literature is concerned with obtaining exponential bounds which are in a sense best possible. However, 'elementary' results such as the Azuma-Hoeffding inequality in Exercise E14.1 are very useful in numerous applications. See for example the applications to combinatorics in Bollobás (1987).

14.10. A consequence of Hölder's inequality

Look at the statement of Doob's $\mathcal{L}^p$ inequality in the next section in order to see where we are going.

LEMMA

Suppose that X and Y are non-negative RVs such that

$$c\mathbf{P}(X \geq c) \leq \mathbf{E}(Y; X \geq c) \text{ for every } c > 0.$$

Then, for $p > 1$ and $p^{-1} + q^{-1} = 1$, we have

$$\|X\|_p \leq q\|Y\|_p.$$

Proof. We obviously have

$$(*) \qquad L := \int_{c=0}^\infty pc^{p-1}\mathbf{P}(X \geq c)dc \leq \int_{c=0}^\infty pc^{p-2}\mathbf{E}(Y; X \geq c)dc =: R.$$

Using Fubini's Theorem with non-negative integrands, we obtain

$$L = \int_{c=0}^{\infty} \left(\int_{\Omega} I_{\{X \geq c\}}(\omega) \mathbf{P}(d\omega) \right) pc^{p-1} dc$$

$$= \int_{\Omega} \left(\int_{c=0}^{X(\omega)} pc^{p-1} dc \right) \mathbf{P}(d\omega) = \mathbf{E}(X^p).$$

Exactly similarly, we find that

$$R = \mathbf{E}(qX^{p-1}Y).$$

We apply Hölder's inequality to conclude that

$$(**) \qquad \mathbf{E}(X^p) \leq \mathbf{E}(qX^{p-1}Y) \leq q\|Y\|_p\|X^{p-1}\|_q.$$

Suppose that $\|Y\|_p < \infty$, *and suppose for now that* $\|X\|_p < \infty$ *also. Then since* $(p-1)q = p$, *we have*

$$\|X^{p-1}\|_q = \mathbf{E}(X^p)^{\frac{1}{q}},$$

so $(**)$ implies that $\|X\|_p \leq q\|Y\|_p$. For general X, note that the hypothesis remains true for $X \wedge n$. Hence $\|X \wedge n\|_p \leq q\|Y\|_p$ for all n, and the result follows using (MON). □

14.11. Doob's $\mathcal{L}^p$ inequality

THEOREM

▶▶(a)   *Let* $p > 1$ *and define* q *so that* $p^{-1} + q^{-1} = 1$. *Let* Z *be a non-negative submartingale bounded in* $\mathcal{L}^p$, *and define (this is standard notation)*

$$Z^* := \sup_{k \in \mathbf{Z}^+} Z_k.$$

Then $Z^* \in \mathcal{L}^p$, *and indeed*

$$(*) \qquad \|Z^*\|_p \leq q \sup_r \|Z_r\|_p.$$

The submartingale Z *is therefore* **dominated** *by the element* Z^* *of* $\mathcal{L}^p$. *As* $n \to \infty$, $Z_\infty := \lim Z_n$ *exists a.s. and in* $\mathcal{L}^p$ *and*

$$\|Z_\infty\|_p = \sup_r \|Z_r\|_p = \uparrow \lim_r \|Z_r\|_p.$$

▶▶(b)   *If Z is of the form $|M|$, where M is a martingale bounded in $\mathcal{L}^p$, then $M_\infty := \lim M_n$ exists a.s. and in $\mathcal{L}^p$, and of course $Z_\infty = |M_\infty|$, a.s.*

Proof. For $n \in \mathbb{Z}^+$, define $Z_n^* := \sup_{k \le n} Z_k$. From Doob's Submartingale Inequality 14.6(a) and Lemma 14.10 we see that

$$\|Z_n^*\|_p \le q\|Z_n\|_p \le q \sup_r \|Z_r\|_p.$$

Property $(*)$ now follows from the Monotone-Convergence Theorem. Since $(-Z)$ is a supermartingale bounded in $\mathcal{L}^p$, and therefore in $\mathcal{L}^1$, we know that $Z_\infty := \lim Z_n$ exists a.s. However,

$$|Z_n - Z_\infty|^p \le (2Z^*)^p \in \mathcal{L}^p,$$

so that (DOM) shows that $Z_n \to Z_\infty$ in $\mathcal{L}^p$. Jensen's inequality shows that $\|Z_r\|_p$ is non-decreasing in r, and all the rest is straightforward.   □

14.12. Kakutani's theorem on 'product' martingales

Let $X_1, X_2, \ldots$ be independent non-negative RVs, each of mean 1. Define $M_0 := 1$, and, for $n \in \mathbb{N}$, let

$$M_n := X_1 X_2 \ldots X_n.$$

Then M is a non-negative martingale, so that

$$M_\infty := \lim M_n \quad \text{exists a.s.}$$

The following five statements are equivalent:

(i) $\mathbb{E}(M_\infty) = 1$, (ii) $M_n \to M_\infty$ in $\mathcal{L}^1$; (iii) M is UI;

(iv) $\prod a_n > 0$ where $0 < a_n := \mathbb{E}(X_n^{\frac{1}{2}}) \le 1$,

(v) $\sum(1 - a_n) < \infty$.

If one (then every one) of the above five statements fails to hold, then

$$\mathbb{P}(M_\infty = 0) = 1.$$

Remark. Something of the significance of this theorem is explained in Section 14.17.

Proof. That $a_n \leq 1$ follows from Jensen's inequality. That $a_n > 0$ is obvious.

First, suppose that statement (iv) *holds.* Then define

$$(*) \qquad\qquad N_n = \frac{X_1^{\frac{1}{2}}}{a_1} \frac{X_2^{\frac{1}{2}}}{a_2} \cdots \frac{X_n^{\frac{1}{2}}}{a_n}.$$

Then N is a martingale for the same reason that M is. See (10.4,b). We have

$$\mathsf{E}N_n^2 = 1/(a_1 a_2 \ldots a_n)^2 \leq 1/(\prod a_k)^2 < \infty,$$

so that N is bounded in $\mathcal{L}^2$. By Doob's $\mathcal{L}^2$ inequality,

$$\mathsf{E}\left(\sup_n |M_n|\right) \leq \mathsf{E}\left(\sup_n |N_n|^2\right) \leq 4\sup_n \mathsf{E}|N_n^2|) < \infty,$$

so that M is dominated by $M^* := \sup_n |M_n| \in \mathcal{L}^1$. Hence M is UI and properties (i)-(iii) hold.

Now consider the case when $\prod a_n = 0$. Define N as at $(*)$. Since N is a non-negative martingale, $\lim N_n$ exists a.s. But since $\prod a_n = 0$, we are forced to conclude that $M_\infty = 0$, a.s.

The equivalence of (iv) and (v) is known to us from (4.3). The theorem is proved. $\qquad\qquad\qquad\qquad\qquad\qquad\qquad\qquad\qquad\qquad\qquad\quad$ $\square$

14.13. The Radon-Nikodým theorem

Martingale theory yields an intuitive – and *'constructive'* – proof of the Radon-Nikodým theorem. We are guided by Meyer (1966).

We begin with a special case.

THEOREM

▶▶(I)   *Suppose that $(\Omega, \mathcal{F}, \mathsf{P})$ is a probability triple in which $\mathcal{F}$ is separable in that*

$$\mathcal{F} = \sigma(F_n : n \in \mathsf{N})$$

for some sequence (F_n) of subsets of Ω. Suppose that Q is a finite measure on $(\Omega, \mathcal{F})$ which is absolutely continuous relative to P in that

(a)   *for $F \in \mathcal{F}$, $\mathsf{P}(F) = 0 \Rightarrow \mathsf{Q}(F) = 0$.*

Then there exists X in $\mathcal{L}^1(\Omega, \mathcal{F}, \mathbf{P})$ such that $\mathbf{Q} = X\mathbf{P}$ (see Section 5.14) in that

$$\mathbf{Q}(F) = \int_F X\, d\mathbf{P} = \mathbf{E}(X; F), \qquad \forall F \in \mathcal{F}.$$

The variable X is called a version of the Radon-Nikodým derivative of $\mathbf{Q}$ relative to $\mathbf{P}$ on $(\Omega, \mathcal{F})$. Two such versions agree a.s. We write

$$\frac{d\mathbf{Q}}{d\mathbf{P}} = X \text{ on } \mathcal{F}, \quad \text{a.s.}$$

Remark. Most of the σ-algebras we have encountered are separable. (The σ-algebra of Lebesgue-measurable subsets of $[0,1]$ is not.)

Proof. With the method of Section 13.1(a) in mind, you can prove that property (a) implies that

(b)   *given $\varepsilon > 0$, there exists $\delta > 0$ such that, for $F \in \mathcal{F}$,*

$$\mathbf{P}(F) < \delta \Rightarrow \mathbf{Q}(F) < \varepsilon.$$

Next, define

$$\mathcal{F}_n := \sigma(F_1, F_2, \ldots, F_n).$$

Then for each n, $\mathcal{F}_n$ consists of the $2^{r(n)}$ possible unions of 'atoms'

$$A_{n,1}, \ldots, A_{n,r(n)}$$

of $\mathcal{F}_n$, an atom A of $\mathcal{F}_n$ being an element of $\mathcal{F}_n$ such that $\emptyset$ is the only proper subset of A which is again an element of $\mathcal{F}_n$. (Each atom A will have the form

$$H_1 \cap H_2 \cap \ldots \cap H_n,$$

where each H_i is either F_i or F_i^c.)

Define a function $X_n : \Omega \to [0, \infty)$ as follows: *if $\omega \in A_{n,k}$, then*

$$X_n(\omega) := \begin{cases} 0 & \text{if} \quad \mathbf{P}(A_{n,k}) = 0, \\ \mathbf{Q}(A_{n,k})/\mathbf{P}(A_{n,k}) & \text{if} \quad \mathbf{P}(A_{n,k}) > 0. \end{cases}$$

Then $X_n \in \mathcal{L}^1(\Omega, \mathcal{F}_n, \mathbf{P})$ and

(c) $$\mathbf{E}(X_n; F) = \mathbf{Q}(F), \qquad \forall F \in \mathcal{F}_n.$$

The variable X_n is the obvious version of $d\mathbb{Q}/d\mathbb{P}$ on $(\Omega, \mathcal{F}_n)$.

It is obvious from (c) that $X = (X_n : n \in \mathbb{Z}^+)$ is a martingale relative to the filtration $(\mathcal{F}_n : n \in \mathbb{Z}^+)$, and since this martingale is non-negative,

$$X_\infty := \lim X_n, \quad \text{exists, a.s.}$$

Let $\varepsilon > 0$, choose δ as at (a), and let $K \in (0, \infty)$ be such that

$$K^{-1}\mathbb{Q}(\Omega) < \delta.$$

Then

$$\mathbb{P}(X_n > K) \le K^{-1}\mathbb{E}(X_n) = K^{-1}\mathbb{Q}(\Omega) < \delta,$$

so that

$$\mathbb{E}(X_n; X_n > K) = \mathbb{Q}(X_n > K) < \varepsilon.$$

The martingale X is therefore UI, so that

$$X_n \to X_\infty \text{ in } \mathcal{L}^1.$$

It now follows from (c) that the measures

$$F \mapsto \mathbb{E}(X_\infty; F) \quad \text{and} \quad F \mapsto \mathbb{Q}(F)$$

agree on the π-system $\bigcup \mathcal{F}_n$, so that they agree on $\mathcal{F}$. All that remains is the proof of uniqueness, which is now standard for us.   □

Remark. The familiarity of all of the arguments in the above proof emphasizes the close link between the Radon-Nikodým and conditional expectation which is made explicit in Section 14.14. Now for the next part of the theorem ...

(II)   *The assumption that $\mathcal{F}$ is separable can be dropped from Part I.*

Once one has Part II, one can easily extend the result to the case when $\mathbb{P}$ and $\mathbb{Q}$ are σ-finite measures by partitioning Ω into sets on which both are finite.

Proving Part II of the theorem is a piece of 'abstract nonsense' based on the fact that $\mathcal{L}^1$ (or, more strictly, L^1) is a metric space, and in particular on the rôle of sequential convergence in metric spaces. *You might well want to take Part II for granted and skip the remainder of this section.*

Let Sep be the class of all separable sub-σ-algebras of $\mathcal{F}$. Part I shows that for $\mathcal{G} \in$ Sep, there exists $X_\mathcal{G}$ in $\mathcal{L}^1(\Omega, \mathcal{G}, \mathbb{P})$ such that

$$d\mathbb{Q}/d\mathbb{P} = X_\mathcal{G} \, ; \text{ equivalently, } \mathbb{E}(X_\mathcal{G}; G) = \mathbb{Q}(G), G \in \mathcal{G}.$$

We are going to prove that there exists X in $\mathcal{L}^1(\Omega, \mathcal{F}, \mathbf{P})$ such that

(d) $$X_\mathcal{G} \to X \text{ in } \mathcal{L}^1$$

in the sense that given $\varepsilon > 0$, there exists $\mathcal{K}$ in Sep such that

$$\text{if } \mathcal{K} \subseteq \mathcal{G} \in \text{Sep, then } \|X_\mathcal{G} - X\|_1 < \varepsilon.$$

First, we note that it is enough to prove that

(e) $$(X_\mathcal{G} : \mathcal{G} \in \text{Sep}) \text{ is Cauchy in } \mathcal{L}^1$$

in the sense that given $\varepsilon > 0$, there exists $\mathcal{K}$ in Sep such that if $\mathcal{K} \subseteq \mathcal{G}_i \in \text{Sep}$ for $i = 1, 2$, then $\|X_{\mathcal{G}_1} - X_{\mathcal{G}_2}\|_1 < \varepsilon$.

Proof that (e) *implies* (d). Suppose that (e) holds. Choose $\mathcal{K}_n \in \text{Sep}$ such that if $\mathcal{K}_n \subseteq \mathcal{G}_i \in \text{Sep}$ for $i = 1, 2$, then

$$\|X_{\mathcal{G}_1} - X_{\mathcal{G}_2}\|_1 < 2^{-(n+1)}.$$

Let $\mathcal{H}(n) = \sigma(\mathcal{K}_1, \mathcal{K}_2, \ldots, \mathcal{K}_n)$. Then (see the proof of (6.10,a)) the limit $X := \lim X_{\mathcal{H}(n)}$ exists a.s. and in $\mathcal{L}^1$, and indeed,

$$\|X - X_{\mathcal{H}(n)}\|_1 \leq 2^{-n}.$$

Set $X := \limsup X_{\mathcal{H}(n)}$ for definiteness. For any $\mathcal{G} \in \text{Sep}$ with $\mathcal{G} \supseteq \mathcal{H}_n$ we have

$$\|X_\mathcal{G} - X_{\mathcal{H}(n)}\|_1 < 2^{-n}.$$

Result (d) follows. □

Proof of (e). If (e) is false, then (why?!) we can find $\varepsilon_0 > 0$ and a sequence $\mathcal{K}(0) \subseteq \mathcal{K}(1) \subseteq \ldots$ of elements of Sep such that

$$\|X_{\mathcal{K}(n)} - X_{\mathcal{K}(n+1)}\|_1 > \varepsilon_0, \quad \forall n.$$

However, it is easily seen that $(X_{\mathcal{K}(n)})$ is a UI martingale relative to the filtration $(\mathcal{K}(n))$, so that $X_{\mathcal{K}(n)}$ converges in $\mathcal{L}^1$. The contradiction establishes that (e) is true. □

Proof of Part II *of the theorem.* We need only show that for X as at (d) and for $F \in \mathcal{F}$, we have

$$\mathbf{E}(X; F) = \mathbf{Q}(F).$$

Choose $\mathcal{K}$ such that for $\mathcal{K} \subseteq \mathcal{G} \in \text{Sep}$, $\|X_{\mathcal{G}} - X\|_1 < \varepsilon$. Then $\sigma(\mathcal{K}, F) \in \text{Sep}$, where $\sigma(\mathcal{K}, F)$ is the smallest σ-algebra extending $\mathcal{K}$ and including F; and, by a familiar argument,

$$|\mathsf{E}(X; F) - \mathbf{Q}(F)| = |\mathsf{E}(X - X_{\sigma(\mathcal{K},F)}; F)|$$
$$\leq \|X - X_{\sigma(\mathcal{K},F)}\|_1 < \varepsilon.$$

The result follows. □

14.14. The Radon-Nikodým theorem and conditional expectation

Suppose that $(\Omega, \mathcal{F}, \mathbf{P})$ is a probability triple, and that $\mathcal{G}$ is a sub-σ-algebra of $\mathcal{F}$. Let X be a non-negative element of $\mathcal{L}^1(\Omega, \mathcal{F}, \mathbf{P})$. Then

$$\mathbf{Q}(X) := \mathsf{E}(X; G), \quad G \in \mathcal{G},$$

defines a finite measure $\mathbf{Q}$ on $(\Omega, \mathcal{G})$. Moreover, $\mathbf{Q}$ is clearly absolutely continuous relative to $\mathbf{P}$ on $\mathcal{G}$, so that, by the Radon-Nikodým theorem, (a version...)

$$Y := d\mathbf{Q}/d\mathbf{P} \text{ exists on } (\Omega, \mathcal{G}).$$

Now Y is $\mathcal{G}$-measurable, and

$$\mathsf{E}(Y; G) = \mathbf{Q}(G) = \mathsf{E}(X; G), \quad G \in \mathcal{G}.$$

Hence Y is a version of the conditional expectation of X given $\mathcal{G}$:

$$Y = \mathsf{E}(X|\mathcal{G}), \quad \text{a.s.}$$

Remark. The right context for appreciating the close inter-relations between martingale convergence, conditional expectation, the Radon-Nikodým theorem, etc., is *the geometry of Banach spaces*.

14.15. Likelihood ratio, equivalent measures

Let $\mathbf{P}$ and $\mathbf{Q}$ be probability measures on $(\Omega, \mathcal{F})$ such that $\mathbf{Q}$ is absolutely continuous relative to $\mathbf{P}$, so that a version X of $d\mathbf{Q}/d\mathbf{P}$ on $\mathcal{F}$ exists. We say that Y is *(a version of) the likelihood ratio of $\mathbf{Q}$ given $\mathbf{P}$*. Then $\mathbf{P}$ is absolutely continuous relative to $\mathbf{Q}$ if and only if $\mathbf{P}(X > 0) = 1$, and then X^{-1} is a version of $d\mathbf{P}/d\mathbf{Q}$. When each of $\mathbf{P}$ and $\mathbf{Q}$ is absolutely continuous relative to the other, then $\mathbf{P}$ and $\mathbf{Q}$ are said to be **equivalent**. Note that it then makes sense to define

$$\int_F \sqrt{d\mathbf{P}d\mathbf{Q}} := \int_F X^{\frac{1}{2}} d\mathbf{P} = \int_F (X^{-\frac{1}{2}}) d\mathbf{Q}, \quad F \in \mathcal{F};$$

and we can hope for a fuller understanding of what Kakutani achieved....

14.16. Likelihood ratio and conditional expectation

Let $(\Omega, \mathcal{F}, \mathbf{P})$ be a probability triple, and let $\mathbf{Q}$ be a probability measure on $(\Omega, \mathcal{F})$ which is absolutely continuous relative to $\mathbf{P}$ with density X. Let $\mathcal{G}$ be a sub-σ-algebra of $\mathcal{F}$. What $\mathcal{G}$-measurable function (modulo versions) yields $d\mathbf{Q}/d\mathbf{P}$ on $\mathcal{G}$? Yes, of course, it is $Y = \mathbf{E}(X|\mathcal{G})$, for, yet again, with $\mathbf{E}$ denoting $\mathbf{P}$-expectation,

$$\mathbf{E}(Y; G) = \mathbf{E}(X; G) = \mathbf{Q}(G) \text{ for } G \in \mathcal{G}.$$

Hence, if $\{\mathcal{F}_n\}$ is a filtration of $(\Omega, \mathcal{F})$, then the likelihood ratios

$$(*) \qquad\qquad (d\mathbf{Q}/d\mathbf{P} \text{ on } \mathcal{F}_n) = \mathbf{E}(X|\mathcal{F}_n)$$

form a UI martingale. (This is of course why the martingale proof of the Radon-Nikodým theorem was bound to succeed!) Here and in the next two sections, we are dropping the 'a.s.' qualifications on such statements as $(*)$: we have outgrown them.

14.17. Kakutani's Theorem revisited; consistency of LR test

Let $\Omega = \mathbf{R}^\mathbf{N}$, $X_n(\omega) = \omega_n$, and define the σ-algebras

$$\mathcal{F} = \sigma(X_k : k \in \mathbf{N}), \qquad \mathcal{F}_n = \sigma(X_k : 1 \le k \le n).$$

Suppose that for each n, f_n and g_n are everywhere positive probability density functions on $\mathbf{R}$ and let $r_n(x) := g_n(x)/f_n(x)$. Let $\mathbf{P}$ [respectively, $\mathbf{Q}$] be the unique measure on $(\Omega, \mathcal{F})$ which makes the variables X_n independent, X_n having probability density function f_n [respectively, g_n]. Clearly, but *you* should prove this,

$$M_n := d\mathbf{Q}/d\mathbf{P} = Y_1 Y_2 \dots Y_n \text{ on } \mathcal{F}_n,$$

where $Y_n = r_n(X_n)$. Note that the variables $(Y_n : n \in \mathbf{N})$ are independent under $\mathbf{P}$ and that each has $\mathbf{P}$-mean 1. For any of a multitude of familiar reasons, M is a martingale.

Now if $\mathbf{Q}$ is absolutely continuous relative to $\mathbf{P}$ on $\mathcal{F}$ with $d\mathbf{Q}/d\mathbf{P} = \xi$ on $\mathcal{F}$, then $M_n = \mathbf{E}(\xi|\mathcal{F}_n)$, and M is UI. Conversely, if M is UI, then M_∞ exists (a.s., $\mathbf{P}$) and

$$\mathbf{E}(M_\infty|\mathcal{F}_n) = M_n, \quad \forall n.$$

But then the probability measures

$$F \mapsto \mathbf{Q}(F) \quad \text{and} \quad F \mapsto \mathbf{E}(M_\infty; F)$$

agree on the π-system $\bigcup \mathcal{F}_n$ and so agree on $\mathcal{F}$, so that $M_\infty = d\mathbf{Q}/d\mathbf{P}$ on $\mathcal{F}$. Thus $\mathbf{Q}$ is absolutely continuous relative to $\mathbf{P}$ on $\mathcal{F}$ if and only if M is UI.

Kakutani's Theorem therefore implies that $\mathbb{Q}$ is equivalent to $\mathbb{P}$ on $\mathcal{F}$ if and only if

$$\prod_n \mathbb{E}(Y_n^{\frac{1}{2}}) = \prod_n \int_{\mathbb{R}} \sqrt{f_n(x)g_n(x)}\,dx > 0,$$

equivalently if

$(*)$ $$\sum_n \int_{\mathbb{R}} \left\{ \sqrt{f_n(x)} - \sqrt{g_n(x)} \right\}^2 dx < \infty;$$

and that then $\mathbb{P}$ is also absolutely continuous relative to $\mathbb{Q}$.

Suppose now that the X_n are *identically distributed* independent variables under each of $\mathbb{P}$ and $\mathbb{Q}$. Thus, for some probability density functions f and g on $\mathbb{R}$, we have $f_n = f$ and $g_n = g$ for all n. It is clear from $(*)$ that $\mathbb{Q}$ is equivalent to $\mathbb{P}$ if and only if $f = g$ almost everywhere with respect to Lebesgue measure, in which case $\mathbb{Q} = \mathbb{P}$. Moreover, Kakutani's Theorem also tells us that if $\mathbb{Q} \neq \mathbb{P}$, then $M_n \to 0$ (a.s., $\mathbb{P}$) and this is exactly *the consistency of the Likelihood-Ratio Test in Statistics.*

14.18. Note on Hardy spaces, etc. (*prestissimo!*)

We have seen in this chapter that for many purposes, the class of UI martingales is a natural one. The appendix to this chapter, on the Optional-Sampling Theorem, provides further evidence of this.

However, what we might wish to be true is not always true for UI martingales. For example, if M is a UI martingale and C is a (uniformly) bounded previsible process, then the martingale $C \bullet M$ need not be bounded in $\mathcal{L}^1$. (Even so, $C \bullet M$ does converge a.s.!)

For many parts of the more advanced theory, one uses the 'Hardy' space $\mathcal{H}_0^1$ of martingales M null at 0 for which one (then each) of the following equivalent conditions holds:

(a) $M^* := \sup |M_n| \in \mathcal{L}^1$,

(b) $[M]_\infty^{\frac{1}{2}} \in \mathcal{L}^1$, where $[M]_n := \sum_{k=1}^n (M_k - M_{k-1})^2$
       and $[M]_\infty =\uparrow \lim[M]_n$.

By a special case of a celebrated **Burkholder-Davis-Gundy theorem**, there exist absolute constants c_p, C_p $(1 \leq p < \infty)$ such that

(c) $c_p \|[M]_\infty^{\frac{1}{2}}\|_p \leq \|M^*\|_p \leq C_p \|[M]_\infty^{\frac{1}{2}}\|_p$ $(1 \leq p < \infty).$

The space $\mathcal{H}_0^1$ is obviously sandwiched between the union of the spaces of martingales bounded in $\mathcal{L}^p$ $(p > 1)$ and the space of UI martingales. Its identification as the right intermediate space has proved very important. Its name derives from its important links with complex analysis.

Proof of the B-D-G inequality or of the equivalence of (a) and (b) is too difficult to give here. But we can take a very quick look at the relevance of (b) to the $C \bullet M$ problem. First, (b) makes it clear that

(d)   *if $M \in \mathcal{H}_0^1$ and C is a bounded previsible process, then $C \bullet M \in \mathcal{H}_0^1$,*

and we shall now see that, in a sense, this is 'best possible'.

Suppose that we have a martingale M null at 0 and a (bounded) previsible process $\varepsilon = (\varepsilon_k : k \in \mathbb{N})$ where the ε_k are IID RVs with $\mathbb{P}(\varepsilon_k = \pm 1) = \frac{1}{2}$, and where ε and M are independent. We want to show that

(e) $M \in \mathcal{H}_0^1$ if (as well as only if) $\varepsilon \bullet M$ is bounded in $\mathcal{L}^1$.

We run into no difficulties of 'regularity' if we condition on M:

$$\mathbb{E}|(\varepsilon \bullet M)_n| = \mathbb{E}\mathbb{E}\{|(\varepsilon \bullet M)_n| \, |\sigma(M)\} \geq 3^{-\frac{1}{2}} \mathbb{E}([M]_n^{\frac{1}{2}}).$$

And where did the last inequality appear from? Let $(a_k : k \in \mathbb{N})$ be a sequence of real numbers. Think of a_k as $M_k - M_{k-1}$ when M is known. Define

$$X_k := a_k \varepsilon_k, \quad W_n := X_1 + \cdots + X_n, \quad v_n = \mathbb{E}(W_n^2) = \sum_{k=1}^{n} a_k^2.$$

Then (see Section 7.2)

$$\mathbb{E}(W_n^4) = \mathbb{E}\left(\sum X_i^4 + 6 \sum\sum_{i<j} X_i^2 X_j^2\right) = \sum_i a_i^4 + 6 \sum\sum_{i<j} a_i^2 a_j^2,$$

so that, certainly, $\mathbb{E}(W_n^4) \leq 3v_n^2$. On combining this fact with Hölder's inequality in the form

$$v_n = \mathbb{E}(W_n^2) \leq \|W^{\frac{2}{3}}\|_{\frac{3}{2}} \|W^{\frac{4}{3}}\|_3 = (\mathbb{E}|W_n|)^{\frac{2}{3}} \mathbb{E}(W_n^4)^{\frac{1}{3}}$$

we obtain the special case of **Khinchine's inequality** we need:

$$\mathbb{E}(|W_n|) \geq 3^{-\frac{1}{2}} v_n^{\frac{1}{2}}. \qquad \qquad \square$$

For more on the topics in this section, see Chow and Teicher (1978), Dellacherie and Meyer (1980), Doob (1981), Durrett (1984). The first of these is accessible to the reader of this book; the others are more advanced.

Chapter 15
Applications

15.0. Introduction – please read!

The purpose of this chapter is to give some indication of some of the ways in which the theory which we have developed can be applied to real-world problems. We consider only very simple examples, but at a lively pace!

In Sections 15.1-15.2, we discuss a trivial case of a celebrated result from mathematical economics, the **Black-Scholes option-pricing formula**. The formula was developed for a continuous-parameter (diffusion) model for stock prices; see, for example, Karatzas and Schreve (1988). We present an obvious discretization which also has many treatments in the literature. What needs to be emphasized is that in the discrete case, the result has nothing to do with probability, which is why the answer is completely independent of the underlying probability measure. *The use of the 'martingale measure' P in Section 15.2 is nothing other than a device for expressing some simple algebra/combinatorics.* But in the diffusion setting, where the algebra and combinatorics are no longer meaningful, the *martingale-representation theorem* and *Cameron-Martin-Girsanov change-of-measure theorem* provide the essential language. I think that this justifies my giving a 'martingale' treatment of something which needs only junior-school algebra.

Sections 15.3-15.5 indicate the further development of the **martingale formulation of optimality in stochastic control,** at which Exercise E10.2 gave a first look. We consider just one 'fun' example, the '*Mabinogion sheep problem*'; but it is an example which illustrates rather well several techniques which may be effectively utilized in other contexts.

In Sections 15.6-15.9, we look at some simple problems of *filtering*: estimating in real-time processes of which only noisy observations can be made. This topic has important applications in engineering (look at the IEEE journals!), in medicine, and in economics. I hope that you will look further into this topic and into the important subject which develops when filtering is combined with stochastic-control theory. See, for example, Davis and Vintner (1985) and Whittle (1990).

Sections 15.10-15.12 consist of first reflections on the problems we encounter when we try to extend the martingale concept.

15.1. A trivial martingale-representation result

Let S denote the two-point set $\{-1, 1\}$, let Σ denote the set of all subsets of S, let $p \in (0, 1)$, and let μ be the probability measure on (S, Σ) with

$$\mu(\{1\}) = p = 1 - \mu(\{-1\}).$$

Let $N \in \mathbf{N}$. Define $(\Omega, \mathcal{F}, \mathbf{P}) = (S, \Sigma, \mu)^N$ so that a typical element of Ω is

$$\omega = (\omega_1, \omega_2, \ldots, \omega_N), \qquad \omega_k \in \{-1, 1\}.$$

Define $\varepsilon_k : \Omega \to \mathbf{R}$ by $\varepsilon_k(\omega) := \omega_k$, so that $(\varepsilon_1, \varepsilon_2, \ldots, \varepsilon_N)$ are IID RVs each with law μ. For $0 \leq n \leq N$, define

$$Z_n := \sum_{k=1}^{n} (\varepsilon_k - 2p + 1),$$
$$\mathcal{F}_n := \sigma(Z_0, Z_1, \ldots, Z_n) = \sigma(\varepsilon_1, \varepsilon_2, \ldots, \varepsilon_n).$$

Note that $\mathsf{E}(\varepsilon_k) = 1.p + (-1)(1 - p) = 2p - 1$. We see that

(a) $$Z = (Z_n : 0 \leq n \leq N)$$

is a martingale (relative to $(\{\mathcal{F}_n : 0 \leq n \leq N\}, \mathbf{P}))$.

LEMMA

If $M = (M_n : 0 \leq n \leq N)$ is a martingale (relative to $(\{\mathcal{F}_n : 0 \leq n \leq N\}, \mathbf{P}))$, then there exists a unique previsible process H such that

$$M = M_0 + H \bullet Z, \text{ that is, } M_n = M_0 + \sum_{k=1}^{n} H_k(Z_k - Z_{k-1}).$$

Remark. Since $\mathcal{F}_0 = \{\emptyset, \Omega\}$, M_0 is constant on Ω and has to be the common value of the $\mathsf{E}(M_n)$.

Proof. We simply construct H explicitly. Because M_n is $\mathcal{F}_n$ measurable,

$$M_n(\omega) = f_n(\varepsilon_1(\omega), \ldots, \varepsilon_n(\omega)) = f_n(\omega_1, \ldots, \omega_n)$$

for some function $f_n : \{-1,1\}^n \to \mathbf{R}$. Since M is a martingale, we have

$$
\begin{aligned}
0 &= \mathsf{E}(M_n - M_{n-1}|\mathcal{F}_{n-1})(\omega) \\
&= pf_n(\omega_1, \ldots, \omega_{n-1}, 1) + (1-p)f_n(\omega_1, \ldots, \omega_{n-1}, -1) \\
&\quad - f_{n-1}(\omega_1, \ldots, \omega_{n-1}).
\end{aligned}
$$

Hence the expressions

(b1)
$$
\frac{f_n(\omega_1, \ldots, \omega_{n-1}, 1) - f_{n-1}(\omega_1, \ldots, \omega_{n-1})}{2(1-p)}
$$

and

(b2)
$$
\frac{f_{n-1}(\omega_1, \ldots, \omega_{n-1}) - f_n(\omega_1, \ldots, \omega_{n-1}, -1)}{2p}
$$

are equal; and if we define $H_n(\omega)$ to be their common value, then H is clearly previsible, and simple algebra verifies that $M = M_0 + H \bullet Z$, as required. *You* check that H is unique.   □

15.2. Option pricing; discrete Black-Scholes formula

Consider an economy in which there are two 'securities': *bonds* of fixed interest rate r, and *stock*, the value of which fluctuates randomly. Let N be a fixed element of $\mathbf{N}$. We suppose that values of units of stock and of bond units change abruptly at times $1, 2, \ldots, N$. For $n = 0, 1, \ldots, \mathbf{N}$, we write

$B_n = (1+r)^n B_0$ for the value of 1 bond unit throughout the open time interval $(n, n+1)$,

S_n for the value of 1 unit of stock throughout the open time interval $(n, n+1)$.

You start just after time 0 with a fortune of value x made up of A_0 units of stock and V_0 of bond, so that

$$
A_0 S_0 + V_0 B_0 = x.
$$

Between times 0 and 1 you invest this in stocks and bonds, so that just before time 1, you have A_1 units of stock and V_1 of bond so that

$$
A_1 S_0 + V_1 B_0 = x.
$$

So, (A_1, V_1) represents the portfolio you have as your 'stake on the first game'.

Just after time $n-1$ (where $n \geq 1$) you have A_{n-1} units of stock and V_{n-1} units of bond with value

$$X_{n-1} = A_{n-1}S_{n-1} + V_{n-1}B_{n-1}.$$

By trading stock for bonds or conversely, you rearrange your portfolio between times $n-1$ and n so that just before time n, your fortune (still of value X_{n-1} because we assume transaction costs to be zero) is described by

$$X_{n-1} = A_n S_{n-1} + V_n B_{n-1} \quad (n \geq 1).$$

Your fortune just after time n is given by

(a) $$X_n = A_n S_n + V_n B_n \quad (n \geq 0)$$

and your change in fortune satisfies

(b) $$X_n - X_{n-1} = A_n(S_n - S_{n-1}) + V_n(B_n - B_{n-1}).$$

Now,

$$B_n - B_{n-1} = rB_n,$$

and

$$S_n - S_{n-1} = R_n S_{n-1},$$

where R_n is the random 'rate of interest of stock at time n'. We may now rewrite (b) as

$$X_n - X_{n-1} = rX_{n-1} + A_n S_{n-1}(R_n - r),$$

so that if we set

(c) $$Y_n = (1+r)^{-n} X_n,$$

then

(d) $$Y_n - Y_{n-1} = (1+r)^{-n} A_n S_{n-1}(R_n - r).$$

Note that (c) shows Y_n to be the discounted value of your fortune at time n, so that the evolution (d) is of primary interest.

Let $\Omega, \mathcal{F}, \varepsilon_n(1 \leq n \leq N)$, $Z_n(0 \leq n \leq N)$ and $\mathcal{F}_n(0 \leq n \leq N)$ be as in Section 15.1. *Note that no probability measure has been introduced.*

We build a model in which each R_n takes only values a, b in $(-1, \infty)$, where

$$a < r < b,$$

by setting

(e)
$$R_n = \frac{a+b}{2} + \frac{b-a}{2}\varepsilon_n.$$

But then

(f) $R_n - r = \frac{1}{2}(b-a)(\varepsilon_n - 2p + 1) = \frac{1}{2}(b-a)(Z_n - Z_{n-1}),$

where we now choose

(g)
$$p := \frac{r-a}{b-a}.$$

Note that (d) and (f) together display Y as a 'stochastic integral' relative to Z.

A *European option* is a contract made just after time 0 which will allow you to buy 1 unit of stock just after time N at a price K; K is the so-called *striking price*. If you have made such a contract, then just after time N, you will exercise the option if $S_N > K$ and will not if $S_N < K$. Thus, the value at time N of such a contract is $(S_N - K)^+$. *What should you pay for the option at time 0?*

Black and Scholes provide an answer to this question which is based on the concept of a *hedging strategy*.

A **hedging strategy with initial value** x for the described option is a portfolio management scheme $\{(A_n, V_n) : 1 \leq n \leq N\}$ where the processes A and V are *previsible* relative to $\{\mathcal{F}_n\}$, and where, with X satisfying (a) and (b), we have **for every** ω,

(h1) $X_0(\omega) = x,$

(h2) $X_n(\omega) \geq 0 \ (0 \leq n \leq N),$

(h3) $X_N(\omega) = (S_N(\omega) - K)^+.$

Anyone employing a hedging strategy will by appropriate portfolio management, and without going bankrupt, exactly duplicate the value of the option at time N.

Note. Though Black and Scholes insist that $X_n(\omega) \geq 0$, $\forall n, \forall \omega$, they (and we) do not insist that the processes A and V be positive. A negative value of V for some n amounts to *borrowing* at the fixed interest rate r. A negative value of A corresponds to '*short-selling*' stock, but after you have read the theorem, this may not worry you!

THEOREM

A hedging strategy with initial value x exists if and only if

$$x = x_0 := \mathsf{E}\left[(1+r)^{-N}(S_N - K)^+\right],$$

where E is the expectation for the measure P of Section 15.1 with p as at (g). There is a unique hedging strategy with initial value x_0, and it involves no short-selling: A is never negative.

On the basis of this result, it is claimed that x_0 is the unique fair price at time 0 of the option.

Proof. In the definition of hedging strategy, there is nowhere any mention of an underlying probability measure. Because however of the 'for every ω' requirement, we should consider only measures on Ω for which each point ω has positive mass. Of course, P is such a measure.

Suppose now that a hedging strategy with initial value x exists, and let A, V, X, Y denote the associated processes. From (d) and (f),

$$Y = Y_0 + F \bullet Z,$$

where F is the *previsible* process with

$$F_n = (1+r)^{-n} A_n S_{n-1}.$$

Of course, F is *bounded* because there are only finitely many (n, ω) combinations. Thus Y is a *martingale* under the P measure, since Z is; and since $Y_0 = x$ and $Y_N = (1+r)^{-N}(S_n - K)^+$ by (c) and the definition of hedging strategy, we obtain

$$x = x_0.$$

(We did not use the property that $X \geq 0$.)

Now consider things afresh and *define*

$$Y_n := \mathsf{E}\left((1+r)^{-N}(S_N - K)^+ | \mathcal{F}_n\right).$$

Then Y is a martingale, and by combining (f) with the martingale-representation result in Section 15.1, we see that for some unique previsible process A, (d) holds. Define

$$X_n := (1+r)^n Y_n, \quad V_n := (X_n - A_n S_n)/B_n.$$

Then (a) and (b) hold. Since

$$X_0 = x \text{ and } X_N = (S_N - K)^+,$$

the only thing which remains is to prove that A is never negative. Because of the explicit formula (15.1,b1), this reduces to showing that

$$\mathsf{E}\left[(S_N - K)^+|S_{n-1}, S_n = (1 + b)S_{n-1}\right]$$
$$\geq \mathsf{E}\left[(S_N - K)^+|S_{n-1}, S_n = (1 + a)S_{n-1}\right];$$

and this is intuitively obvious and may be proved by a simple computation on binomial coefficients.

15.3. The Mabinogion sheep problem

In the Tale of Peredur ap Efrawg in the very early Welsh folk tales, *The Mabinogion* (see Jones and Jones (1949)), there is a magical flock of sheep, some black, some white. We sacrifice poetry for precision in specifying its behaviour. At each of times $1, 2, 3, \ldots$ a sheep (chosen randomly from the entire flock, independently of previous events) bleats; if this bleating sheep is white, one black sheep (if any remain) instantly becomes white; if the bleating sheep is black, then one white sheep (if any remain) instantly becomes black. No births or deaths occur.

The controlled system

Suppose now that the system can be controlled in that just after time 0 and just after every magical transition time $1, 2, \ldots$, any number of white sheep may be removed from the system. (White sheep may be removed on numerous occasions.) The object is to maximize the expected final number of black sheep.

Consider the following example of a policy:

> *Policy A*: at each time of decision, do nothing if there are more black sheep than white sheep or if no black sheep remain; otherwise, immediately reduce the white population to one less than the black population.

The *value function V* for Policy A is the function

$$V : \mathsf{Z}^+ \times \mathsf{Z}^+ \to [0, \infty),$$

where for $w, b \in \mathsf{Z}^+$, $V(w, b)$ denotes the expected final number of black sheep if one adopts Policy A and if there are w white and b black sheep at time 0. Then V is uniquely specified by the fact that, for $w, b \in \mathsf{Z}^+$,

(a1) $V(0, b) = b$;

(a2) $V(w,b) = V(w-1,b)$ whenever $w \geq b$ and $w > 0$;

(a3) $V(w,b) = \frac{w}{w+b}V(w+1,b-1) + \frac{b}{w+b}V(w-1,b+1)$ whenever $w < b$, $b > 0$ and $w > 0$.

It is almost tautological that if W_n and B_n denote the numbers of white and black sheep at time n, then, if we adopt Policy A, then, whatever the initial values of W_0 and B_0,

(b) $V(W_n, B_n)$ *is a martingale*

relative to the natural filtration of $\{(W_n, B_n) : n \geq 0\}$.

(c) LEMMA

The following statements are true for $w, b \in \mathbf{Z}^+$:

(c1) $V(w,b) \geq V(w-1,b)$ *whenever* $w > 0$,

(c2) $V(w,b) \geq \frac{w}{w+b}V(w+1,b-1) + \frac{b}{w+b}V(w-1,b+1)$
whenever $w > 0$ *and* $b > 0$.

Let us suppose that this Lemma is true. (It is proved in the next Section.) Then, for any policy whatsoever,

(d) $V(W_n, B_n)$ *is a supermartingale.*

The fact that $V(W_n, B_n)$ converges means that the system must a.s. end up in an absorbing state in which sheep are of one colour. But then $V(W_\infty, B_\infty)$ is just the final number of black sheep (by definition of V). Since $V(W_n, B_n)$ is a non-negative supermartingale, we have for deterministic W_0, B_0,

$$EV(W_\infty, B_\infty) \leq V(W_0, B_0).$$

Hence, whatever the initial position, the expected final number of black sheep under *any* policy is no more than the expected final number of black sheep if Policy A is used. Thus

Policy A is optimal.

In Section 15.5, we prove the following result:

(e) $V(k,k) - (2k + \frac{\pi}{4} - \sqrt{\pi k}) \to 0$ as $k \to \infty$.

Thus if we start with 10 000 black sheep and 10 000 white sheep, we finish up with (about) 19 824 black sheep on average (over many 'runs').

Of course, the above argument worked because we had correctly guessed the optimal policy to begin with. In this subject area, one often has to make good guesses. Then one usually has to work rather hard to clinch results which correspond in more general situations to Lemma (c) and statement (d). You might find it quite an amusing exercise to prove these results for our special problem now, before reading on.

For a problem in economics which utilizes analogous ideas, see Davis and Norman (1990).

15.4. Proof of Lemma 15.3(c)

It will be convenient to define

(a) $$v_k := V(k, k).$$

Everything hinges on the following results: for $1 \leq c \leq k$,

(b1) $$V(k - c, k + c) = v_k + (2k - v_k)2^{-(2k-2)} \sum_{j=k}^{k+c-1} \binom{2k-1}{j},$$

(b2) $V(k + 1 - c, k + c)$

$$= v_k + (2k + 1 - v_k)\left\{2^{2k-1} + \tfrac{1}{2}\binom{2k}{k}\right\}^{-1}\left\{\sum_{j=k}^{k+c-1} \binom{2k}{j}\right\},$$

which simply reflect (15.3,a3) together with the 'boundary conditions':

$$V(k, k) = v_k, \quad V(0, 2k) = 2k,$$
(c)
$$V(k + 1, k) = v_k, \quad V(0, 2k + 1) = 2k + 1.$$

Now, from (15.3,a2),

$$v_{k+1} = V(k + 1, k + 1) = V(k, k + 1),$$

and hence, from (b2) with $c = 1$, we find that

(d) $$v_{k+1} = \frac{1 - p_k}{1 + p_k}v_k + \frac{2p_k}{1 + p_k}(2k + 1),$$

where p_k is the chance of obtaining k heads and k tails in $2k$ tosses of a fair coin:

(e) $$p_k = 2^{-2k}\binom{2k}{k}.$$

Result (d) is the key to proving things by induction.

Proof of result (15.3,c1). From (15.3,a2), result (15.3,c1) is automatically true when $w \geq b$. Hence, we need only establish the result when $w < b$. Now if $w < b$ and $w + b$ is odd, then

$$(w, b) = (k + 1 - c, k + c)$$

for some c with $1 \leq c \leq k$. But formulae (b) show that it is enough to show that for $1 \leq a \leq k$,

$$(2k + 1 - v_k) \left\{ 2^{2k-1} + \frac{1}{2} \binom{2k}{k} \right\}^{-1} \binom{2k}{k + a - 1}$$
$$\geq (2k - v_k) 2^{-(2k-2)} \binom{2k - 1}{k + a - 1};$$

and since

$$\binom{2k}{k + a - 1} \bigg/ \binom{2k}{k} \geq \binom{2k - 1}{k + a - 1} \bigg/ \binom{2k - 1}{k},$$

we need only establish the case when $a = 1$:

$$(2k + 1 - v_k) 2^{-(2k-1)} (1 + p_k)^{-1} \binom{2k}{k}$$
$$\geq (2k - v_k) 2^{-(2k-2)} \binom{2k - 1}{k},$$

which reduces to

(f) $$v_k \geq 2k - p_k^{-1}.$$

But property (f) follows by induction from (d) using only the fact that

$$p_k \text{ is decreasing in } k.$$

Proof for the case when $b + w$ is even may be achieved similarly.

Proof of result (15.3,c2). Because of (15.3,a3), the result (15.3,c2) is automatically true when $w < b$, so we need only establish it when $w \geq b$. In analogy with the reduction of the 'general a' case to the 'boundary case $a = 1$' in the proof of (15.3,c1), it is easily shown that it is sufficient to prove (15.3,c2) for the case when $(w, b) = (k + 1, k + 1)$ for some k. Formulae (b) reduce this problem to showing that

$$\{1 + (2k + 1)p_k\} v_k \leq 2k(2k + 1)p_k,$$

and this follows by induction from (d) using only the fact that

$$(2k+1)p_k \text{ is increasing in } k. \qquad \square$$

15.5. Proof of result (15.3,e)

Define

$$\alpha_k := v_k - 2k - (p_k)^{-1} - \tfrac{1}{4}\pi.$$

Then, from (15.4,d),

$$\alpha_{k+1} = (1 - \rho_k)\alpha_k + \rho_k c_k,$$

where

$$\rho_k := \frac{2p_k}{1 + p_k}, \quad c_k := \frac{p_k - p_{k+1}}{2p_k^2 p_{k+1}} - \frac{\pi}{4}.$$

Stirling's formula shows that $c_k \to 0$ as $k \to \infty$, so that given $\varepsilon > 0$, we can find N so that

$$|c_k| < \varepsilon \quad \text{for} \quad k \geq N.$$

Induction shows that for $k \geq N$,

$$|\alpha_{k+1}| \leq (1 - \rho_k)(1 - \rho_{k-1})\ldots(1 - \rho_N)|\alpha_N| + \varepsilon.$$

But, since $\sum \rho_k = \infty$, we have $\prod(1 - \rho_k) = 0$, and it is now clear that $\limsup |\alpha_{k+1}| \leq \varepsilon$, so that $\alpha_k \to 0$.

Because of the accurate version of Stirling's formula:

$$n! = (2\pi n)^{\frac{1}{2}} \left(\frac{n}{e}\right)^n e^{\theta/(12n)}, \quad 0 < \theta = \theta(n) < 1,$$

we have

$$p_k^{-1} = (\pi k)^{\frac{1}{2}} \left\{1 + O\left(\frac{1}{k}\right)\right\},$$

so that

$$v_k - \left(2k + \frac{\pi}{4} - \sqrt{\pi k}\right) \to 0,$$

as required. $\qquad \square$

We now take a quick look at filtering. The central idea combines Bayes' formula with a recursive property which is now illustrated by two examples.

15.6. Recursive nature of conditional probabilities

Example. Suppose that A, B, C and D are events (elements of $\mathcal{F}$) each with a strictly positive probability. Let us write (for example) ABC for $A \cap B \cap C$. Let us also introduce the notation

$$\mathcal{C}_A(B) := \mathbf{P}(B|A) = \mathbf{P}(AB)/\mathbf{P}(A)$$

for conditional probabilities. The 'recursive property' in which we are interested is exemplified by

$$\mathcal{C}_{ABC}(D) = \mathcal{C}_{AB}(D|C) := \frac{\mathcal{C}_{AB}(CD)}{\mathcal{C}_{AB}(C)} \; ;$$

'if we want to find the conditional probability of D given that A, B and C have occurred, we can assume that both A and B have occurred and find the $\mathcal{C}_{AB}$ probability of D given C'. □

Example. Suppose the X, Y, Z and T are RVs such that (X, Y, Z, T) has a strictly positive joint pdf $f_{X,Y,Z,T}$ on $\mathbf{R}^4$: for $B \in \mathcal{B}^4$,

$$\mathbf{P}\{(X, Y, Z, T) \in B\} = \int \int \int \int_B f_{X,Y,Z,T}(x, y, z, t)dx\,dy\,dz\,dt.$$

Then, of course, (X, Y, Z) has joint pdf $f_{X,Y,Z}$ on $\mathbf{R}^3$, where

$$f_{X,Y,Z}(x, y, z) = \int_{\mathbf{R}} f_{X,Y,Z,T}(x, y, z, t)dt.$$

The formula

$$f_{T|X,Y,Z}(t|x, y, z) := \frac{f_{X,Y,Z,T}(x, y, z, t)}{f_{X,Y,Z}(x, y, z)}$$

defines a ('regular') conditional pdf of T given X, Y, Z: for $B \in \mathcal{B}$, we have, with all dependence on ω indicated,

$$\mathbf{P}(T \in B|X, Y, Z)(\omega) = \mathbf{E}(I_B(T)|X, Y, Z)(\omega)$$
$$= \int_B f_{T|X,Y,Z}(t|X(\omega), Y(\omega), Z(\omega))dt.$$

Similarly,

$$f_{T,Z|X,Y}(t, z|x, y) = \frac{f_{X,Y,Z,T}(x, y, z, t)}{f_{X,Y}(x, y)} .$$

The recurrence property is exemplified by

$$f_{T|X,Y,Z} = (f_{T|Z})_{|X,Y} := \frac{f_{T,Z|X,Y}}{f_{Z|X,Y}}. \qquad \square$$

15.7. Bayes' formula for bivariate normal distributions

With a now-clear notation, we have for RVs X, Y with strictly positive joint pdf $f_{X,Y}$ on $\mathbf{R}^2$,

$$(*) \qquad f_{X|Y}(x|y) = \frac{f_{X,Y}(x,y)}{f_Y(y)} = \frac{f_X(x)f_{Y|X}(y|x)}{f_Y(y)}.$$

Thus

$$(**) \qquad f_{X|Y}(x|y) \propto f_X(x)f_{Y|X}(y|x),$$

the 'constant of proportionality' depending on y but being determined by the fact that

$$\int_{\mathbf{R}} f_{X|Y}(x|y)dx = 1.$$

The meaning of the following Lemma is clarified within its proof.

LEMMA

▶(a)   *Suppose that $\mu, a, b \in \mathbf{R}$, that $U, W \in (0, \infty)$ and that X and Y are RVs such that*
$$\mathcal{L}(X) = \mathrm{N}(\mu, U),$$
$$\mathcal{C}_X(Y) = \mathrm{N}(a + bX, W).$$

   Then
$$\mathcal{C}_Y(X) = \mathrm{N}(\hat{X}, V),$$

   where the number $V \in (0, \infty)$ and the RV $\hat{X}$ are determined as follows:
$$\frac{1}{V} = \frac{1}{U} + \frac{b^2}{W}, \quad \frac{\hat{X}}{V} = \frac{\mu}{U} + \frac{b(Y-a)}{W}.$$

Proof. The absolute distribution of X is $\mathrm{N}(\mu, U)$, so that

$$f_X(x) = (2\pi U)^{-\frac{1}{2}} \exp\left\{-\frac{(x-\mu)^2}{2U}\right\}.$$

The conditional pdf of Y given X is the density of $N(a + bX, W)$, so that

$$f_{Y|X}(y|x) = (2\pi W)^{-\frac{1}{2}} \exp\left\{-\frac{(y - a - bx)^2}{2W}\right\}.$$

Hence, from (**),

$$\log f_{X|Y}(x|y) = c_1(y) - \frac{(x - \mu)^2}{2U} - \frac{(y - a - bx)^2}{2W}$$

$$= c_2(y) - \frac{(x - \hat{x})^2}{2V},$$

where $1/V = 1/U + b^2/W$ and $\hat{x}/V = \mu/U + b(y - a)/W$. The result follows.
□

COROLLARY

(b)   *With the notation of the Lemma, we have*

$$\|X - \hat{X}\|_2^2 = E\{(X - \hat{X})^2\} = V.$$

Proof. Since $C_Y(X) = N(\hat{X}, V)$, we even have

$$E\{(X - \hat{X})^2|Y\} = V, \text{ a.s.} \qquad\qquad □$$

15.8. Noisy observation of a single random variable

Let $X, \eta_1, \eta_2, \dots$ be independent RVs, with

$$\mathcal{L}(X) = N(0, \sigma^2), \quad \mathcal{L}(\eta_k) = N(0, 1).$$

Let (c_n) be a sequence of positive real numbers, and let

$$Y_k = X + c_k\eta_k, \quad \mathcal{F}_n = \sigma(Y_1, Y_2, \dots, Y_n).$$

We regard each Y_k as a noisy observation of X. We know that

$$M_n := E(X|\mathcal{F}_n) \to M_\infty := E(X|\mathcal{F}_\infty) \text{ a.s. and in } \mathcal{L}^2.$$

One interesting question mentioned at (10.4,c) is:

> *when is it true that* $M_\infty = X$ *a.s.?, or again,*

> *is* X *a.s. equal to an* $\mathcal{F}_\infty$*-measurable RV?*

Let us write $\mathcal{C}_n$ to signify 'regular conditional law' given $Y_1, Y_2, \ldots, Y_n$. We have

$$\mathcal{C}_0(X) = \mathrm{N}(0, \sigma^2).$$

Suppose that it is true that

$$\mathcal{C}_{n-1}(X) = \mathrm{N}(\hat{X}_{n-1}, V_{n-1})$$

where $\hat{X}_{n-1}$ is a linear function of $Y_1, Y_2, \ldots, Y_{n-1}$ and V_{n-1} a constant in $(0, \infty)$. Then, since $Y_n = X + c_n \eta_n$, we have

$$\mathcal{C}_{n-1}(Y_n | X) = \mathrm{N}(X, c_n^2).$$

From the Lemma 15.7(a) on bivariate normals with

$$\mu = \hat{X}_{n-1}, U = V_{n-1}, a = 0, b = 1, W = c_n^2,$$

we have

$$\mathcal{C}_{n-1}(X | Y_n) = \mathrm{N}(\hat{X}_n, V_n),$$

where

$$\frac{1}{V_n} = \frac{1}{V_{n-1}} + \frac{1}{c_n^2}, \quad \frac{\hat{X}_n}{V_n} = \frac{\hat{X}_{n-1}}{V_n} + \frac{Y_n}{c_n^2}.$$

But the recursive property indicated in Section 15.6 shows that

$$\mathcal{C}_n(X) = \mathcal{C}_{n-1}(X | Y_n).$$

We have now proved by induction that

$$\mathcal{C}_n(X) = \mathrm{N}(\hat{X}_n, V_n), \quad \forall n.$$

Now, of course,

$$M_n = \hat{X}_n \quad \text{and} \quad \mathrm{E}\{(X - M_n)^2\} = V_n.$$

However,

$$V_n = \{\sigma^{-2} + \sum_{k=1}^{n} c_k^{-2}\}^{-1}.$$

Our martingale M is $\mathcal{L}^2$ bounded and so converges in $\mathcal{L}^2$. We now see that

$$M_\infty = X \ a.s. \ if \ and \ only \ if \ \sum c_k^{-2} = \infty .$$

15.9. The Kalman-Bucy filter

The method of calculation used in the preceding three sections allows immediate derivation of the famous Kalman-Bucy filter.

Let A, H, C, K and g be real constants. Suppose that $X_0, Y_0, \varepsilon_1, \varepsilon_2, \ldots,$ $\eta_1, \eta_2, \ldots$ are independent RVs with

$$\mathcal{L}(\varepsilon_k) = \mathcal{L}(\eta_k) = \mathrm{N}(0,1), \quad \mathcal{L}(X_0) = \mathrm{N}(m, \sigma^2), Y_0 = 0.$$

The true state of a system at time n is supposed given by X_n, where

(**dynamics**)
$$X_n - X_{n-1} = AX_{n-1} + H\varepsilon_n + g.$$

However the process X cannot be observed directly: we can only observe the process Y, where

(**observation**)
$$Y_n - Y_{n-1} = CX_n + K\eta_n.$$

Just as in Section 15.8, we make the induction hypothesis that

$$\mathcal{C}_{n-1}(X_{n-1}) = \mathrm{N}(\hat{X}_{n-1}, V_{n-1}),$$

where $\mathcal{C}_{n-1}$ signifies regular conditional law given $Y_1, Y_2, \ldots, Y_{n-1}$. Since $X_n = \alpha X_{n-1} + g + H\varepsilon_n$, where

$$\alpha := 1 + A,$$

we have

$$\mathcal{C}_{n-1}(X_n) = \mathrm{N}(\alpha \hat{X}_{n-1} + g, \alpha^2 V_{n-1} + H^2).$$

Also, since $Y_n = Y_{n-1} + CX_n + K\eta_n$, we have

$$\mathcal{C}_{n-1}(Y_n | X_n) = \mathrm{N}(Y_{n-1} + CX_n, K^2).$$

Apply the bivariate-normal Lemma 15.7(a) to find that

$$\mathcal{C}_n(X_n) = \mathcal{C}_{n-1}(X_n | Y_n) = \mathrm{N}(\hat{X}_n, V_n),$$

where

(KB1)
$$\frac{1}{V_n} = \frac{1}{\alpha^2 V_{n-1} + H^2} + \frac{C^2}{K^2},$$

(KB2)
$$\frac{\hat{X}_n}{V_n} = \frac{\alpha \hat{X}_{n-1} + g}{\alpha^2 V_{n-1} + H^2} + \frac{C(Y_n - Y_{n-1})}{K^2}.$$

Equation (KB1) shows that $V_n = f(V_{n-1})$. Examination of the rectangular hyperbola which is the graph of f shows that $V_n \to V_\infty$, the positive root of $V = f(V)$.

If one wishes to give a rigorous treatment of the K-B filter in continuous time, one is forced to use martingale and stochastic-integral techniques. See, for example, Rogers and Williams (1987) and references to filtering and control mentioned there. For more on the discrete-time situation, which is very important in practice, and for how filtering does link with stochastic control, see Davis and Vintner (1985) or Whittle (1990).

15.10. Harnesses entangled

The martingale concept is well adapted to processes evolving in time because (discrete) time naturally belongs to the *ordered* space $\mathbf{Z}^+$. The question arises: does the martingale property transfer in some natural way to processes parametrized by (say) $\mathbf{Z}^d$?

Let me first explain a difficulty described in Williams (1973) that arises with models in $\mathbf{Z}$ $(d = 1)$ and in $\mathbf{Z}^2$, though we do not study the latter here.

Suppose that $(X_n : n \in \mathbf{Z})$ is a process such that each $X_n \in \mathcal{L}^1$ and that ('almost surely' qualifications will be dropped)

(a) $\qquad E(X_n|X_m : m \neq n) = \tfrac{1}{2}(X_{n-1} + X_{n+1}), \quad \forall n.$

For $m \in \mathbf{Z}$, define

$$\mathcal{G}_m = \sigma(X_k : k \leq m), \quad \mathcal{H}_m = \sigma(X_r : r \geq m).$$

The Tower Property shows that for a, b in $\mathbf{Z}^+$ with $a < b$, we have for $a < r < b$,

$$E(X_r|\mathcal{G}_a, \mathcal{H}_b) = E(X_r|X_s : r \neq s|\mathcal{G}_a, \mathcal{H}_b)$$
$$= \tfrac{1}{2}E(X_{r-1}|\mathcal{G}_a, \mathcal{H}_b) + \tfrac{1}{2}E(X_{r+1}|\mathcal{G}_a, \mathcal{H}_b),$$

so that $r \mapsto E(X_r|\mathcal{G}_a, \mathcal{H}_b)$ is the linear interpolation

$$E(X_r|\mathcal{G}_a, \mathcal{H}_b) = \frac{b - r}{b - a} X_a + \frac{r - a}{b - a} X_b.$$

Hence, for $n \in \mathbf{Z}$ and $u \in \mathbf{N}$, we have, a.s.,

$$E(X_n|\mathcal{G}_{n-u}, \mathcal{H}_{n+1}) = \frac{X_{n-u}}{u + 1} + \frac{u X_{n+1}}{u + 1}.$$

Now, the σ-algebras $\sigma(\mathcal{G}_{n-u}, \mathcal{H}_{n+1})$ decrease as $u \uparrow \infty$. Because of Warning 4.12, we had better not claim that they decrease to $\sigma(\mathcal{G}_{-\infty}, \mathcal{H}_{n+1})$. Anyway, by the Downward Theorem, we see that the random variable

$$L := \lim_{u \downarrow -\infty}(X_u/u) \text{ exists (a.s.)}$$

and

$$\mathsf{E}(X_n | \bigcap_u \sigma(\mathcal{G}_{n-u}, \mathcal{H}_{n+1})) = X_{n+1} - L.$$

Hence, by the Tower Property,

(b) $$\mathsf{E}(X_n | L, \mathcal{H}_{n+1}) = X_{n+1} - L$$

whence we have a reversed-martingale property:

$$\mathsf{E}(X_n - nL | L, \mathcal{H}_{n+1}) = X_{n+1} - (n+1)L.$$

A further application of the Downward Theorem shows that

(c) $$A := \lim_{n \uparrow \infty}(X_n - nL) \text{ exists (a.s.)}$$

Hence

$$L = \lim_{u \uparrow \infty}(X_u/u).$$

By using the arguments which led to (b) in the reversed-time sense, we now obtain

(d) $$\mathsf{E}(X_{n+1} | L, \mathcal{G}_n) = X_n + L.$$

From (b) and (d) and the Tower Property,

$$\mathsf{E}(X_n + L | X_{n+1}) = X_{n+1},$$
$$\mathsf{E}(X_{n+1} | X_n + L) = X_n + L.$$

Exercise E9.2 shows that $X_{n+1} = X_n + L$. Hence (almost surely)

$$X_n = nL + A, \quad \forall n \in \mathbf{Z},$$

so that (almost) *all sample paths of X are straight lines!*

Hammersley (1966) suggested that any analogue of (a) should be called a *harness* property and that the type of result just obtained conveys the idea that every low-dimensional harness is a straitjacket!

15.11. Harnesses unravelled, 1

The reason that (15.10,a) rules out interesting models is that it is expressed in terms of the idea that each X_n should be a random variable. What one should say is that

$$X_n : \Omega \to \mathbf{R}$$

but then require only that *differences*

$$(X_r - X_s : r, s \in \mathbf{Z})$$

be RVs (that is, be $\mathcal{F}$-measurable), and that for $n, k \in \mathbf{Z}$ with $k \neq n$,

$$\mathsf{E}(X_n - X_k | X_m - X_k : m \neq n) = \tfrac{1}{2}(X_{n-1} - X_k) + \tfrac{1}{2}(X_{n+1} - X_k).$$

I call this a *difference harness* in Williams (1973).

Easy exercise. Suppose that $(Y_n : n \in \mathbf{Z})$ are IID RVs in $\mathcal{L}^1$. Let X_0 be any function on Ω. Define

$$X_n := \begin{cases} X_0 + \sum_{k=1}^n Y_k & \text{if } n \geq 0, \\ X_0 - \sum_{k=n+1}^0 Y_k & \text{if } n < 0. \end{cases}$$

Thus, $X_n - X_{n-1} = Y_n, \forall n$. Prove that X is a difference harness.

15.12. Harnesses unravelled, 2

In dimension $d \geq 3$, we do not need to use the 'difference-process' unravelling described in the preceding section. For $d \geq 3$, there is a non-trivial model, related both to *Gibbs states* in statistical mechanics and to *quantum fields*, such that each $X_n(n \in \mathbf{Z}^d)$ *is* a RV and, for $n \in \mathbf{Z}^d$,

$$\mathsf{E}(X_n | X_m : m \in \mathbf{Z}^d \setminus \{n\}) = (2d)^{-1} \sum_{u \in \mathcal{U}} X_{n+u},$$

where $\mathcal{U}$ is the set of the $2d$ unit vectors in $\mathbf{Z}^d$. See Williams (1973).

In addition to a fascinating etymological treatise on the terms 'martingale' and 'harness', Hammersley (1966) contains many important ideas on harnesses, anticipating later work on *stochastic partial differential equations*. Many interesting unsolved problems on various kinds of harness remain.

PART C: CHARACTERISTIC FUNCTIONS

Chapter 16
Basic Properties of CFs

Part C is merely the briefest account of the first stages of characteristic function theory. This theory is something very different in spirit from the work in Part B. Part B was about the sample paths of processes. Characteristic function theory is on the one hand part of Fourier-integral theory, and it is proper that it finds its way into that marvellous recent book, Körner (1988); see also the magical Dym and McKean (1972). On the other hand, characteristic functions do have an essential rôle in both probability and statistics, and I must include these few pages on them. For full treatment, see Chow and Teicher (1978) or Lukacs (1970).

Exercises indicate the analogous Laplace-transform method, and develop in full the method of moments for distributions on [0,1].

16.1. Definition

The *characteristic function* (CF) $\varphi = \varphi_X$ of a random variable X is defined to be the map

$$\varphi : \mathbf{R} \to \mathbf{C}$$

(**important**:the domain is $\mathbf{R}$ not $\mathbf{C}$) defined by

▶▶ $$\varphi(\theta) := \mathsf{E}(e^{i\theta X}) = \mathsf{E}\cos(\theta X) + i\mathsf{E}\sin(\theta X).$$

Let $F := F_X$ be the distribution function of X, and let $\mu := \mu_X$ denote the law of X. Then

$$\varphi(\theta) := \int_{\mathbf{R}} e^{i\theta x} dF(x) := \int_{\mathbf{R}} e^{i\theta x} \mu(dx),$$

so that φ is the Fourier transform of μ, or the Fourier-Stieltjes transform of F. (We do not use the factor $(2\pi)^{-\frac{1}{2}}$ which is sometimes used in Fourier theory.) We often write φ_F or φ_μ for φ.

16.2. Elementary properties

Let $\varphi = \varphi_X$ for a RV X. Then

►(a) $\varphi(0) = 1$ (obviously);

►(b) $|\varphi(\theta)| \leq 1$, $\forall \theta$;

►(c) $\theta \mapsto \varphi(\theta)$ is continuous on **R**;

(d) $\varphi_{(-X)}(\theta) = \overline{\varphi_X(\theta)}$, $\forall \theta$,

(e) $\varphi_{aX+b}(\theta) = e^{ib\theta}\varphi_X(a\theta)$.

You can easily prove these properties. (Use (BDD) to establish (c).)

Note on differentiability (or lack of it).   Standard analysis (see Theorem A16.1) implies that if $n \in \mathbf{N}$ and $E(|X|^n) < \infty$, then we may formally differentiate $\varphi(\theta) = Ee^{i\theta X}$ n times to obtain

$$\varphi^{(n)}(\theta) = E[(iX)^n e^{i\theta X}].$$

In particular, $\varphi^{(n)}(0) = i^n E(X^n)$. However, it is possible for $\varphi'(0)$ to exist when $E(|X|) = \infty$.

We shall see shortly that φ can be the 'tent-function'

$$\varphi(\theta) = (1 - |\theta|)I_{[-1,1]}(\theta)$$

so that φ need not be differentiable everywhere, and φ can be 0 outside $[-1, 1]$.

16.3. Some uses of characteristic functions

Amongst uses of CFs are the following:

- to prove the Central Limit Theorem (CLT) and analogues,
- to calculate distributions of limit RVs,
- to prove the 'only if' part of the Three-Series Theorem,
- to obtain estimates on tail probabilities via saddle-point approximation,
- to prove such results as
  if X and Y are independent, and $X + Y$ has a normal distribution, then both X and Y have normal distributions.

Only the first three of these uses are discussed in this book.

16.4. Three key results

(a) *If X and Y are independent RVs, then*

$$\varphi_{X+Y}(\theta) = \varphi_X(\theta)\varphi_Y(\theta), \quad \forall \theta.$$

Proof. This is just 'independence means multiply' again:

$$\mathsf{E}e^{i\theta(X+Y)} = \mathsf{E}e^{i\theta X}e^{i\theta Y} = \mathsf{E}e^{i\theta X}\mathsf{E}e^{i\theta Y}. \qquad \square$$

(b) *F may be reconstructed from φ.*

See Section 16.6 for a precise statement.

(c) *'Weak' convergence of distribution functions corresponds exactly to convergence of the corresponding CFs.*

See Section 18.1 for a precise statement.

The way in which these results are used in the proof of the Central Limit Theorem is as follows. Suppose that $X_1, X_2, \ldots$ are IID RVs each with mean 0 and variance 1. From (a) and (16.2,e), we see that if $S_n := X_1 + \cdots + X_n$, then

$$\mathsf{E}\exp(i\theta S_n/\sqrt{n}) = \varphi_X(\theta/\sqrt{n})^n.$$

We shall obtain rigorous estimates which show that

$$\varphi_X(\theta/\sqrt{n})^n = \left\{1 - \frac{1}{2}\theta^2/n + o(1/n)\right\}^n \to \exp(-\frac{1}{2}\theta^2), \quad \forall \theta.$$

Since $\theta \mapsto \exp(-\frac{1}{2}\theta^2)$ is the CF of the standard normal distribution (as we shall see shortly), it now follows from (b) and (c) that

the distribution of $S_n/\sqrt{n}$ converges weakly to the distribution function Φ of the standard normal distribution $N(0,1)$.

In this case, this simply means that

$$\mathsf{P}(S_n/\sqrt{n} \leq x) \to \Phi(x), \quad x \in \mathbf{R}.$$

16.5. Atoms

In regard to both (16.4,b) and (16.4,c), tidiness of results can be threatened by the presence of atoms.

If $P(X = c) > 0$, then the law μ of X is said to have an *atom* at c, and the distribution function F of X has a discontinuity at c:

$$\mu(\{c\}) = F(c) - F(c-) = P(X = c).$$

Now μ can have at most n atoms of mass at least $1/n$, so that the number of atoms of μ is at most countable.

It therefore follows that given $x \in \mathbf{R}$, there exists a sequence (y_n) of reals with $y_n \downarrow x$ such that every y_n is a non-atom of μ (equivalently a point of continuity of F); and then, by right-continuity of F, $F(y_n) \downarrow F(x)$.

16.6. Lévy's Inversion Formula

This theorem puts the fact that F may be reconstructed from φ in very explicit form. (Check that the theorem does imply that if F and G are distribution functions such that $\varphi_F = \varphi_G$ on $\mathbf{R}$, then $F = G$.)

THEOREM

▶   *Let φ be the CF of a RV X which has law μ and distribution function F. Then, for $a < b$,*

(a)
$$\lim_{T \uparrow \infty} \frac{1}{2\pi} \int_{-T}^{T} \frac{e^{-i\theta a} - e^{-i\theta b}}{i\theta} \varphi(\theta) d\theta$$
$$= \frac{1}{2}\mu(\{a\}) + \mu(a, b) + \frac{1}{2}\mu(\{b\})$$
$$= \frac{1}{2}[F(b) + F(b-)] - \frac{1}{2}[F(a) + F(a-)].$$

Moreover, if $\int_{\mathbf{R}} |\varphi(\theta)| d\theta < \infty$, then X has continuous probability density function f, and

(b)
$$f(x) = \frac{1}{2\pi} \int_{\mathbf{R}} e^{-i\theta x} \varphi(\theta) d\theta.$$

The 'duality' between (b) and the result

(c)
$$\varphi(\theta) = \int_{\mathbf{R}} e^{i\theta x} f(x) dx$$

can be exploited as we shall see.

The proof of the theorem may be omitted on a first reading.

Proof of the theorem. For $u, v \in \mathbf{R}$ with $u \le v$,

(d) $$|e^{iv} - e^{iu}| \le |v - u|,$$

either from a picture or since

$$\left| \int_u^v i e^{it} dt \right| \le \int_u^v |i e^{it}| dt = \int_u^v 1 dt.$$

Let $a, b \in \mathbf{R}$, with $a < b$. Fubini's Theorem allows us to say that, for $0 < T < \infty$,

(e)
$$\frac{1}{2\pi} \int_{-T}^T \frac{e^{-i\theta a} - e^{-i\theta b}}{i\theta} \varphi(\theta) d\theta$$

$$= \frac{1}{2\pi} \int_{-T}^T \frac{e^{-i\theta a} - e^{-i\theta b}}{i\theta} \left(\int_{\mathbf{R}} e^{i\theta x} \mu(dx) \right) d\theta$$

$$= \frac{1}{2\pi} \int_{\mathbf{R}} \left\{ \int_{-T}^T \frac{e^{i\theta(x-a)} - e^{i\theta(x-b)}}{i\theta} d\theta \right\} \mu(dx)$$

provided we show that

$$C_T := \frac{1}{2\pi} \int_{\mathbf{R}} \left\{ \int_{-T}^T \left| \frac{e^{i\theta(x-a)} - e^{i\theta(x-b)}}{i\theta} \right| d\theta \right\} \mu(dx) < \infty.$$

However, inequality (d) shows that $C_T \le (b-a)T/\pi$, so that (e) is valid.

Next, we can exploit the evenness of the cosine function and the oddness of the sine function to obtain

(f)
$$\frac{1}{2\pi} \int_{-T}^T \frac{e^{i\theta(x-a)} - e^{i\theta(x-b)}}{i\theta} d\theta$$

$$= \frac{\operatorname{sgn}(x-a) S(|x-a|T) - \operatorname{sgn}(x-b) S(|x-b|T)}{\pi},$$

where, as usual,

$$\operatorname{sgn}(x) := \begin{cases} 1 & \text{if } x > 0, \\ 0 & \text{if } x = 0, \\ -1 & \text{if } x < 0, \end{cases}$$

and

$$S(U) := \int_0^U \frac{\sin x}{x} dx \qquad (U > 0).$$

Even though the Lebesgue integral $\int_0^\infty x^{-1} \sin x\, dx$ does not exist, because

$$\int_0^\infty \left(\frac{\sin x}{x}\right)^+ dx = \int_0^\infty \left(\frac{\sin x}{x}\right)^- dx = \infty,$$

we have (see Exercise E16.1)

$$\lim_{U\uparrow\infty} S(U) = \frac{\pi}{2}.$$

The expression (f) is bounded simultaneously in x and T for our fixed a and b; and, as $T \uparrow \infty$, the expression (f) converges to

0 if $x < a$ or $x > b$,
$\frac{1}{2}$ if $x = a$ or $x = b$,
1 if $a < x < b$.

The **Bounded Convergence Theorem** now yields result (a).

Suppose now that $\int_{\mathbf{R}} |\varphi(\theta)| d\theta < \infty$. We can then let $T \uparrow \infty$ in result (a) and use (DOM) to obtain

(g) $$F(b) - F(a) = \frac{1}{2\pi} \int_{\mathbf{R}} \frac{e^{-i\theta a} - e^{-i\theta b}}{i\theta} \varphi(\theta) d\theta,$$

provided that F is continuous at a and b. However, (DOM) shows that the right-hand side of (g) is continuous in a and b and (why?!) we can conclude that F has no atoms and that (g) holds for all a and b with $a < b$.

We now have

(h) $$\frac{F(b) - F(a)}{b - a} = \frac{1}{2\pi} \int_{\mathbf{R}} \frac{e^{-i\theta a} - e^{-i\theta b}}{i\theta(b - a)} \varphi(\theta) d\theta.$$

But, by (d),

$$\left| \frac{e^{-i\theta a} - e^{-i\theta b}}{i\theta(b - a)} \right| \le 1.$$

Hence, the assumption that $\int_{\mathbf{R}} |\varphi(\theta)| d\theta < \infty$ allows us to use (DOM) to let $b \to a$ in (h) to obtain

$$F'(a) = f(a) := \frac{1}{2\pi} \int_{\mathbf{R}} e^{-i\theta a} \varphi(\theta) d\theta,$$

and, finally, f is continuous by (DOM). $\qquad\qquad\square$

16.7. A table

	Distribution	pdf	Support	CF
1.	$N(\mu, \sigma^2)$	$\frac{1}{\sigma\sqrt{2\pi}}\exp\left\{-\frac{(x-\mu)^2}{2\sigma^2}\right\}$	**R**	$\exp(i\mu\theta - \frac{1}{2}\sigma^2\theta^2)$
2.	$U[0,1]$	1	[0,1]	$\frac{e^{i\theta}-1}{i\theta}$
3.	$U[-1,1]$	$\frac{1}{2}$	[-1,1]	$\frac{\sin\theta}{\theta}$
4.	Double exponential	$\frac{1}{2}e^{-|x|}$	**R**	$\frac{1}{1+\theta^2}$
5.	Cauchy	$\frac{1}{\pi(1+x^2)}$	**R**	$e^{-|\theta|}$
6.	Triangular	$1-|x|$	[-1,1]	$2\left(\frac{1-\cos\theta}{\theta^2}\right)$
7.	Anon	$\frac{1-\cos x}{\pi x^2}$	**R**	$(1-|\theta|)I_{[-1,1]}(\theta)$

The two lines 4 and 5 illustrate the duality between (16.6,b) and (16.6,c), as do the two lines 6 and 7. Hints on verifying the table are given in the exercises on this chapter.

Chapter 17
Weak Convergence

In this chapter, we consider the appropriate concept of 'convergence' for probability measures on $(\mathbf{R}, \mathcal{B})$. The terminology 'weak convergence' is unfortunate: the concept is closer to 'weak*' than to 'weak' convergence in the senses used by functional analysts. 'Narrow convergence' is the official pure-mathematical term. However, probabilists seem determined to use 'weak convergence' in their sense, and, reluctantly, I follow them here.

We are studying the special case of weak convergence on a Polish (complete, separable, metric) space S when $S = \mathbf{R}$; and we unashamedly use special features of $\mathbf{R}$. For the general theory, see Billingsley (1968) or Parthasarathy (1967) – or, for a superb acount of its current scope, Ethier and Kurtz (1986).

Notation. We write

$$\mathrm{Prob}(\mathbf{R})$$

for the space of probability measures on $\mathbf{R}$, and

$$C_b(\mathbf{R})$$

for the space of bounded continuous functions on $\mathbf{R}$.

17.1. The 'elegant' definition

Let $(\mu_n : n \in \mathsf{N})$ be a sequence in $\mathrm{Prob}(\mathbf{R})$ and let $\mu \in \mathrm{Prob}(\mathbf{R})$. We say that μ_n **converges weakly to** μ if (and only if)

(a) $$\mu_n(h) \to \mu(h), \qquad \forall h \in C_b(\mathbf{R}),$$

and then write

(b) $$\mu_n \xrightarrow{w} \mu$$

We know that elements of $\mathrm{Prob}(\mathbb{R})$ correspond to distribution functions via the correspondence

$$\mu \leftrightarrow F, \qquad \text{where} \quad F(x) = \mu(-\infty, x].$$

Weak convergence of distribution functions is defined in the obvious way:

(c) $\qquad\qquad F_n \xrightarrow{w} F$ if and only if $\mu_n \xrightarrow{w} \mu.$

We are generally interested in the case when $F_n = F_{X_n}$, that is when F_n is the distribution function for some random variable X_n. Then, by (6.12), we have, for $h \in C_b(\mathbb{R})$,

$$\mu_n(h) = \int_{\mathbb{R}} h(x) dF_n(x) = \mathbb{E}h(X_n).$$

Note that the statement $F_n \xrightarrow{w} F$ is meaningful even if the X_n's are defined on different probability spaces.

However, *if X_n ($n \in \mathbb{N}$) and X are RVs on the same triple $(\Omega, \mathcal{F}, \mathbb{P})$, then*

(d) $\qquad\qquad (X_n \to X, \text{ a.s.}) \quad \Rightarrow \quad (F_{X_n} \xrightarrow{w} F_X),$

and indeed,

(e) $\qquad\qquad (X_n \to X \text{ in prob}) \quad \Rightarrow \quad (F_{X_n} \xrightarrow{w} F_X),$

Proof of (d). Suppose that $X_n \to X$, a.s., and that μ_n is the law of X_n and μ is the law of X. Then, for $h \in C_b(\mathbb{R})$, we have $h(X_n) \to h(X)$, a.s., and, by (BDD),

$$\mu_n(h) = \mathbb{E}h(X_n) \to \mathbb{E}h(X) = \mu(h). \qquad\qquad \square$$

Exercise. Prove (e).

17.2. A 'practical' formulation

Example. Atoms are a nuisance. Suppose that $X_n = \frac{1}{n}$, $X = 0$. Let μ_n be the law of X_n, so that μ_n is the unit mass at $\frac{1}{n}$, and let μ be the law of X. Then, for $h \in C_b(\mathbb{R})$,

$$\mu_n(h) = h(n^{-1}) \to h(0) = \mu(h),$$

so that $\mu_n \xrightarrow{w} \mu$. However,

$$F_n(0) = 0 \nrightarrow F(0) = 1. \qquad\qquad \square$$

LEMMA

(a) Let (F_n) be a sequence of DFs on **R**, and let F be a DF on **R**. Then $F_n \overset{w}{\to} F$ if and only if

$$\lim_n F_n(x) = F(x)$$

for every non-atom (that is, every point of continuity) x of F.

Proof of 'only if' part. Suppose that $F_n \overset{w}{\to} F$. Let $x \in \mathbf{R}$, and let $\delta > 0$. Define $h \in C_b(\mathbf{R})$ via

$$h(y) := \begin{cases} 1 & \text{if } y \le x, \\ 1 - \delta^{-1}(y - x) & \text{if } x < y < x + \delta, \\ 0 & \text{if } y \ge x + \delta. \end{cases}$$

Then $\mu_n(h) \to \mu(h)$. Now,

$$F_n(x) \le \mu_n(h) \quad \text{and} \quad \mu(h) \le F(x + \delta),$$

so that

$$\limsup_n F_n(x) \le F(x + \delta).$$

However, F is right-continuous, so we may let $\delta \downarrow 0$ to obtain

(b) $$\limsup_n F_n(x) \le F(x), \qquad \forall x \in \mathbf{R}.$$

In similar fashion, working with $y \mapsto h(y + \delta)$, we find that for $x \in \mathbf{R}$ and $\delta > 0$,

$$\liminf_n F_n(x-) \ge F(x - \delta),$$

so that

(c) $$\liminf_n F_n(x-) \ge F(x-), \qquad \forall x \in \mathbf{R}.$$

Inequalities (b) and (c) refine the desired result.   □

In the next section, we obtain the 'if' part as a consequence of a nice representation.

17.3. Skorokhod representation

THEOREM

> *Suppose that $(F_n : n \in \mathbf{N})$ is a sequence of DFs on $\mathbf{R}$, that F is a DF on $\mathbf{R}$ and that $F_n(x) \to F(x)$ at every point x of continuity of F.*
>
> *Then there exists a probability triple $(\Omega, \mathcal{F}, \mathbf{P})$ carrying a sequence (X_n) of RVs and also a RV X such that*

$$F_n = F_{X_n}, \quad F = F_X,$$

and

$$X_n \to X \quad \text{a.s.}$$

This is a kind of 'converse' to (17.1,d).

Proof. We simply use the construction in Section 3.12. Thus, take

$$(\Omega, \mathcal{F}, \mathbf{P}) = ([0,1], \mathcal{B}[0,1], \text{Leb}),$$

define

$$X^+(\omega) := \inf\{z : F(z) > \omega\},$$
$$X^-(\omega) := \inf\{z : F(z) \geq \omega\},$$

and define X_n^+, X_n^- similarly. We know from Section 3.12 that X^+ and X^- have DF F and that $\mathbf{P}(X^+ = X^-) = 1$.

Fix ω. Let z be a non-atom of F with $z > X^+(\omega)$. Then $F(z) > \omega$, and hence, for large n, $F_n(z) > \omega$, so that $X_n^+(\omega) \leq z$. So $\limsup_n X_n^+(\omega) \leq z$. But (since non-atoms are dense), we can choose $z \downarrow X^+(\omega)$. Hence

$$\limsup X_n^+(\omega) \leq X^+(\omega),$$

and, by similar arguments,

$$\liminf X_n^-(\omega) \geq X^-(\omega).$$

Since $X_n^- \leq X_n^+$ and $\mathbf{P}(X^+ = X^-) = 1$, the result follows. $\square$

17.4. Sequential compactness for $\text{Prob}(\overline{\mathbf{R}})$

There is a problem in working with the non-compact space $\mathbf{R}$. Let μ_n be the unit mass at n. No subsequence of (μ_n) converges weakly in $\text{Prob}(\mathbf{R})$, but $\mu_n \xrightarrow{w} \mu_\infty$ in $\text{Prob}(\overline{\mathbf{R}})$, where μ_∞ is the unit mass at ∞. Here $\overline{\mathbf{R}}$ is the compact metrizable space $[-\infty, \infty]$, the definition of $\mathrm{Prob}(\overline{\mathbf{R}})$ is obvious, and $\mu_n \overset{w}{\to} \mu_\infty$ in $\mathrm{Prob}(\overline{\mathbf{R}})$ means that

$$\mu_n(h) \to \mu_\infty(h), \qquad \forall h \in C(\overline{\mathbf{R}}).$$

(We do not need the subscript 'b' on $C(\overline{\mathbf{R}})$ because elements of $C(\overline{\mathbf{R}})$ are bounded.) It is important to keep remembering that while functions in $C(\overline{\mathbf{R}})$ tend to limits at $+\infty$ and $-\infty$, functions in $C_b(\mathbf{R})$ need not. The space $C(\overline{\mathbf{R}})$ is separable (it has a countable dense subset) while the space $C_b(\mathbf{R})$ is not.

Let me briefly describe how one should think of the next topic. Here I assume some functional analysis, but from the next paragraph (not the next section) on, I resort to elementary bare-hands treatment. By the Riesz representation theorem, the dual space $C(\overline{\mathbf{R}})^*$ of $C(\overline{\mathbf{R}})$ is the space of bounded signed measures on $(\overline{\mathbf{R}}, \mathcal{B}(\overline{\mathbf{R}}))$. The weak* topology of $C(\overline{\mathbf{R}})^*$ is metrizable (because $C(\overline{\mathbf{R}})$ is separable), and under this topology the unit ball of $C(\overline{\mathbf{R}})^*$ is compact and contains $\mathrm{Prob}(\overline{\mathbf{R}})$ as a closed subset. The weak* topology of $\mathrm{Prob}(\overline{\mathbf{R}})$ is *exactly* the probabilists' weak topology, so

(a) $\mathrm{Prob}(\overline{\mathbf{R}})$ *is a compact metrizable space under our probabilists' weak topology.*

The bare-hands substitute for result (a) is the following.

LEMMA (Helly-Bray)

(b) *Let (F_n) be any sequence of distribution functions on $\mathbf{R}$. Then there exist a right-continuous non-decreasing function F on $\mathbf{R}$ such that $0 \leq F \leq 1$ and a subsequence (n_i) such that*

(*) $\lim_i F_{n_i}(x) = F(x)$ *at every point of continuity F.*

Proof. We make an obvious use of 'the diagonal principle'.

Take a countable dense set C of $\mathbf{R}$ and label it:

$$C = \{c_1, c_2, c_3, \ldots\}.$$

Since $(F_n(c_1) : n \in \mathbf{N})$ is a bounded sequence, it contains a convergent subsequence $(F_{n(1,j)}(c_1))$:

$$F_{n(1,j)}(c_1) \to H(c_1) \text{ (say)} \qquad \text{as } j \to \infty.$$

In some subsequence of *this* subsequence, we shall have

$$F_{n(2,j)}(c_2) \to H(c_2) \qquad \text{as } j \to \infty;$$

and so on. If we put $n_i = n(i,i)$, then we shall have:

$$H(c) := \lim F_{n_i}(c) \text{ exists for every } c \text{ in } C.$$

Obviously, $0 \leq H \leq 1$, and H is non-decreasing on C. For $x \in \mathbb{R}$, define

$$F(x) := \lim_{c \downarrow\downarrow x} H(c),$$

the '$\downarrow\downarrow$' signifying that c decreases *strictly* to x through C. (In particular, $F(c)$ need not equal $H(c)$ for c in C.)

Our function F is right-continuous, as you can check. By the 'limsupery' of Sections 17.2 and 17.3, you can also check *for yourself* that (∗) holds: I wouldn't dream of depriving you of that pleasure.   □

17.5. Tightness

Of course, the function F in the Helly-Bray Lemma 17.4(b) need not be a distribution function. It will be a distribution if and only if

$$\lim_{x \downarrow\downarrow -\infty} F(x) = 0, \qquad \lim_{x \uparrow\uparrow +\infty} F(x) = 1.$$

Definition

▶▶   A sequence (F_n) of distribution functions is called *tight* if, given $\varepsilon > 0$, there exists $K > 0$ such that, for every n,

$$\mu_n[-K, K] = F_n(K) - F_n(-K-) > 1 - \varepsilon.$$

You can see the idea: 'tightness stops mass being pushed out to $+\infty$ or $-\infty$'.

LEMMA

Suppose that F_n is a sequence of DFs.

(a)   If $F_n \overset{w}{\to} F$ for some DF F, then (F_n) is tight.

(b)   If (F_n) is tight, then there exists a subsequence (F_{n_i}) and a DF F such that $F_{n_i} \overset{w}{\to} F$.

This is a really easy limsupery exercise.

Chapter 18
The Central Limit Theorem

The Central Limit Theorem (CLT) is one of the great results of mathematics. Here we derive it as a corollary of **Lévy's Convergence Theorem** which says that *weak convergence of DFs corresponds exactly to pointwise convergence of CFs.*

18.1. Lévy's Convergence Theorem

▶▶▶   *Let (F_n) be a sequence of DFs, and let φ_n denote the CF of F_n. Suppose that*

$$g(\theta) := \lim \varphi_n(\theta) \text{ exists for all } \theta \in \mathbf{R},$$

and that

$$g(\cdot) \text{ is continuous at } 0.$$

Then $g = \varphi_F$ for some distribution function F, and

$$F_n \xrightarrow{w} F.$$

Proof. Assume for the moment that

(a)   *the sequence (F_n) is tight.*

Then, by the Helly-Bray result 17.5(b), we can find a subsequence (F_{n_k}) and a DF F such that

$$F_{n_k} \xrightarrow{w} F.$$

But then, for $\theta \in \mathbf{R}$, we have

$$\varphi_{n_k}(\theta) \to \varphi_F(\theta) \qquad (\varphi_{n_k} : CF \text{ of } F_{n_k})$$

(take $h(x) = e^{i\theta x}$). Thus $g = \varphi_F$.

Now we argue by contradiction. Suppose that (F_n) does not converge weakly to F. Then, for some point x of continuity of F, we can find a subsequence which we shall denote by $(\tilde{F}_n)$ and an $\eta > 0$ such that

$$(*) \qquad\qquad |\tilde{F}_n(x) - F(x)| \geq \eta, \qquad \forall n.$$

But $(\tilde{F}_n)$ is tight, so that we can find a subsequence $\tilde{F}_{n_j}$ and a DF $\tilde{F}$ such that

$$\tilde{F}_{n_j} \xrightarrow{w} \tilde{F}.$$

But then $\tilde{\varphi}_{n_j} \to \tilde{\varphi}$, so that $\tilde{\varphi} = \varphi_{\tilde{F}} = g = \varphi_F$. Since a CF determines the corresponding DF uniquely, we see that $\tilde{F} = F$, so that, in particular, x is a non-atom of $\tilde{F}$ and

$$(**) \qquad\qquad \tilde{F}_{n_j}(x) \to \tilde{F}(x) = F(x).$$

The contradiction between $(*)$ and $(**)$ clinches the result. $\qquad\qquad \square$

We must now prove (a).

Proof of tightness of (F_n). Let $\varepsilon > 0$ be given. Since the expression

$$\varphi_n(\theta) + \varphi_n(-\theta) = \int_{\mathbf{R}} 2\cos(\theta x) dF_n(x)$$

is real, it follows that $g(\theta) + g(-\theta)$ is real (and obviously bounded above by 2).

Since g is continuous at 0 and equal to 1 at 0, we can choose $\delta > 0$ such that

$$|1 - g(\theta)| < \tfrac{1}{4}\varepsilon \quad \text{when} \quad |\theta| < \delta.$$

We now have

$$0 < \delta^{-1} \int_0^{\delta} \{2 - g(\theta) - g(-\theta)\} d\theta \leq \tfrac{1}{2}\varepsilon.$$

Since $g = \lim \varphi_n$, the Bounded Convergence Theorem for the finite interval $[0, \delta]$ shows that there exists n_0 in $\mathbf{N}$ such that for $n \geq n_0$,

$$\delta^{-1} \int_0^{\delta} \{2 - \varphi_n(\theta) - \varphi_n(-\theta)\} d\theta \leq \varepsilon.$$

However,

$$\delta^{-1} \int_0^\delta \{2 - \varphi_n(\theta) - \varphi_n(-\theta)\} d\theta$$

$$= \delta^{-1} \int_{-\delta}^\delta \left\{ \int_{\mathbf{R}} (1 - e^{i\theta x}) \, dF_n(x) \right\} d\theta$$

$$= \int_{\mathbf{R}} \left\{ \delta^{-1} \int_{-\delta}^\delta (1 - e^{i\theta x}) \, d\theta \right\} dF_n(x),$$

the interchange of order of integration being justified by the fact that since $|1 - e^{i\theta x}| \leq 2$, 'the integral of the absolute value' is clearly finite. We now have, for $n \geq n_0$,

$$\varepsilon \geq 2 \int_{\mathbf{R}} \left(1 - \frac{\sin \delta x}{\delta x}\right) dF_n \geq 2 \int_{|x| > 2\delta^{-1}} \left(1 - \frac{1}{|\delta x|}\right) dF_n(x)$$

$$\geq \int_{|x| > 2\delta^{-1}} dF_n = \mu_n\{x : |x| > 2\delta^{-1}\}$$

and it is now evident that the sequence (F_n) is tight.   □

If you now re-read Section 16.4, you will realize that the next task is to obtain 'Taylor' estimates on characteristic functions.

18.2. o and O notation

Recall that

$$f(t) = O(g(t)) \quad \text{as} \quad t \to L$$

means that

$$\limsup_{t \to L} |f(t)/g(t)| < \infty$$

and that

$$f(t) = o(g(t)) \quad \text{as} \quad t \to L$$

means that

$$f(t)/g(t) \to 0 \quad \text{as} \quad t \to L.$$

18.3. Some important estimates

For $n = 0, 1, 2, \ldots$ and x real, define the 'remainder'

$$R_n(x) = e^{ix} - \sum_{k=0}^{n} \frac{(ix)^k}{k!}.$$

Then

$$R_0(x) = e^{ix} - 1 = \int_0^x ie^{iy}dy,$$

and from these two expressions we see that

$$|R_0(x)| \leq \min(2, |x|).$$

Since

$$R_n(x) = \int_0^x iR_{n-1}(y)dy,$$

we obtain by induction:

$$|R_n(x)| \leq \min\left(\frac{2|x|^n}{n!}, \frac{|x|^{n+1}}{(n+1)!}\right).$$

Suppose now that X is a zero-mean RV in $\mathcal{L}^2$:

$$\mathsf{E}(X) = 0, \quad \sigma^2 := \mathrm{Var}(X) < \infty.$$

Then, with φ denoting φ_X, we have

▶(a) $$|\varphi(\theta) - (1 - \tfrac{1}{2}\sigma^2\theta^2)| = |\mathsf{E}R_2(\theta X)| \leq \mathsf{E}|R_2(\theta X)|$$

$$\leq \theta^2 \mathsf{E}\left(|X|^2 \wedge \frac{|\theta||X|^3}{6}\right).$$

The final term within $\mathsf{E}(\cdot)$ is dominated by the integrable RV $|X|^2$ and tends to 0 as $\theta \to 0$. Hence, by (DOM), we have

▶▶(b) $$\varphi(\theta) = 1 - \tfrac{1}{2}\sigma^2\theta^2 + \mathrm{o}(\theta^2) \quad \text{as} \quad \theta \to 0.$$

Next, for $|z| < \tfrac{1}{2}$, and with principal values for logs,

$$\log(1 + z) - z = \int_0^z \frac{(-w)}{1 + w}dw = -z^2 \int_0^1 \frac{tdt}{1 + tz},$$

and since $|1 + tz| \geq \tfrac{1}{2}$ we have

▶▶(c) $$|\log(1 + z) - z| \leq |z|^2, \qquad |z| \leq \tfrac{1}{2}.$$

18.4. The Central Limit Theorem

▶▶▶   *Let (X_n) be an IID sequence, each X_n distributed as X where*

$$\mathsf{E}(X) = 0, \quad \sigma^2 := \mathrm{Var}(X) < \infty.$$

Define $S_n := X_1 + \cdots + X_n$, and set

$$G_n := \frac{S_n}{\sigma\sqrt{n}}.$$

Then, for $x \in \mathbf{R}$, we have, as $n \to \infty$,

$$\mathsf{P}(G_n \le x) \to \Phi(x) = \frac{1}{\sqrt{2\pi}} \int_{-\infty}^{x} \exp(-\tfrac{1}{2}y^2)dy.$$

Proof. Fix θ in $\mathbf{R}$. Then, using (18.3,b),

$$\varphi_{G_n}(\theta) = \varphi_{S_n}\left(\frac{\theta}{\sigma\sqrt{n}}\right) = \varphi\left(\frac{\theta}{\sigma\sqrt{n}}\right)^n$$

$$= \left\{1 - \frac{1}{2}\frac{\theta^2}{n} + \mathrm{o}\left(\frac{\theta^2}{\sigma^2 n}\right)\right\}^n,$$

the 'o' now referring to the situation when $n \to \infty$. But now, using (18.3,c), we have, as $n \to \infty$,

$$\log \varphi_{G_n}(\theta) = n \log\left\{1 - \frac{1}{2}\frac{\theta^2}{n} + \mathrm{o}\left(\frac{\theta^2}{n}\right)\right\}$$

$$= n\left\{-\frac{1}{2}\frac{\theta^2}{n} + \mathrm{o}\left(\frac{\theta^2}{n}\right)\right\} \to -\frac{1}{2}\theta^2.$$

Hence $\varphi_{G_n}(\theta) \to \exp(-\tfrac{1}{2}\theta^2)$, and since $\theta \mapsto \exp(-\tfrac{1}{2}\theta^2)$ is the CF of the normal distribution, the result follows from Theorem 18.1.   □

18.5. Example

Let us look at a simple example which shows how the method may be adapted to deal with a sequence of independent but non-IID RVs.

With the Record Problem E4.3 in mind, suppose that on some $(\Omega, \mathcal{F}, \mathsf{P})$, $E_1, E_2, \ldots$ are independent events with $\mathsf{P}(E_n) = 1/n$. Define

$$N_n = I_{E_1} + \cdots + I_{E_n},$$

the 'number of records by time n' in the record context. Then

$$E(N_n) = \sum_{k \le n} \frac{1}{k} = \log n + \gamma + o(1), \quad (\gamma \text{ is Euler's constant})$$

$$\text{Var}(N_n) = \sum_{k \le n} \frac{1}{k}\left(1 - \frac{1}{k}\right) = \log n + \gamma - \frac{\pi^2}{6} + o(1).$$

Let

$$G_n := \frac{N_n - \log n}{\sqrt{\log n}}$$

so that $E(G_n) \to 0$, $\text{Var}(G_n) \to 1$. Then, for fixed θ in **R**,

$$\varphi_{G_n}(\theta) = \exp(-i\theta\sqrt{\log n})\varphi_{N_n}\left(\frac{\theta}{\sqrt{\log n}}\right).$$

But

$$\varphi_{N_n}(t) = \prod_{k=1}^{n} \varphi_{X_k}(t) = \prod_{k=1}^{n}\left\{1 - \frac{1}{k} + \frac{1}{k}e^{it}\right\}.$$

We see that as $n \to \infty$, and with $t := \theta/\sqrt{\log n}$,

$$\varphi_{G_n}(\theta) = -i\theta\sqrt{\log n} + \sum_{k=1}^{n}\log\left\{1 + \frac{1}{k}\left(e^{it} - 1\right)\right\}$$

$$= -i\theta\sqrt{\log n} + \sum_{k=1}^{n}\log\left\{1 + \frac{1}{k}\left[it - \frac{1}{2}t^2 + o(t^2)\right]\right\}$$

$$= -i\theta\sqrt{\log n} + \sum_{k=1}^{n}\frac{1}{k}\left(it - \frac{1}{2}t^2 + o(t^2)\right) + O\left(\sum_{k=1}^{n}\frac{t^2}{k^2}\right)$$

$$= -i\theta\sqrt{\log n} + \left(it - \frac{1}{2}t^2 + o(t^2)\right)[\log n + O(1)] + t^2 O(1)$$

$$= -\frac{1}{2}\theta^2 + o(1) \to -\frac{1}{2}\theta^2.$$

Hence $P(G_n \le x) \to \Phi(x)$, $x \in \mathbf{R}$.   □

See Hall and Heyde (1980) for some very general limit theorems.

18.6. CF proof of Lemma 12.4

Lemma 12.4 gave us the 'only if' part of the Three-Series Theorem 12.5. Its statement was as follows.

LEMMA

Suppose that (X_n) is a sequence of independent random variables bounded by a constant K in $[0, \infty)$:

$$|X_n(\omega)| \leq K, \quad \forall n, \forall \omega.$$

Then

$$(\textstyle\sum X_n \text{ converges, a.s.}) \Rightarrow (\textstyle\sum \mathbb{E}(X_n) \text{ converges and } \textstyle\sum \mathrm{Var}(X_n) < \infty).$$

The proof given in Section 12.4 was rather sophisticated.

Proof using characteristic functions. First, note that, as a consequence of estimate (18.3,a), if Z is a RV such that for some constant K_1,

$$|Z| \leq K_1, \quad \mathbb{E}(Z) = 0, \quad \sigma^2 := \mathrm{Var}(Z) < \infty,$$

then for $|\theta| < K_1^{-1}$, we have

$$|\varphi_Z(\theta)| \leq 1 - \frac{1}{2}\sigma^2\theta^2 + \frac{1}{6}|\theta|^3 K_1 \mathbb{E}(Z^2)$$

$$\leq 1 - \frac{1}{2}\sigma^2\theta^2 + \frac{1}{6}\sigma^2\theta^2 = 1 - \frac{1}{3}\sigma^2\theta^2$$

$$\leq \exp\left(-\frac{1}{3}\sigma^2\theta^2\right).$$

Now take $Z_n := X_n - \mathbb{E}(X_n)$. Then

$$\mathbb{E}(Z_n) = 0, \quad \mathrm{Var}(Z_n) = \mathrm{Var}(X_n),$$

$$|\varphi_{Z_n}(\theta)| = |\exp\{-i\theta\mathbb{E}(X_n)\}\varphi_{X_n}(\theta)| = |\varphi_{X_n}(\theta)|,$$

and $|Z_n| \leq 2K$.

If $\sum \mathrm{Var}(X_n) = \infty$, then we shall have, for $0 < |\theta| < (2K)^{-1}$,

$$\prod |\varphi_{X_k}(\theta)| = \prod |\varphi_{Z_k}(\theta)| \leq \exp\left\{-\frac{1}{3}\theta^2 \sum \mathrm{Var}(X_k)\right\} = 0.$$

However, if $\sum X_k$ converges a.s. to S, then, by (DOM),

$$\prod_{k \leq n} \varphi_{X_k}(\theta) = \mathbb{E}\exp(i\theta S_n) \to \varphi_S(\theta),$$

and $\varphi_S(\theta)$ is continuous in θ with $\varphi_S(0) = 1$. We have a contradiction.

Hence $\sum \mathrm{Var}(X_n) = \sum \mathrm{Var}(Z_n) < \infty$, and, since $\mathbb{E}(Z_n) = 0$, Theorem 12.2(a) shows that

$$\sum Z_n \quad \text{converges a.s.}$$

Hence

$$\sum \mathbb{E}(X_n) = \sum \{X_n - Z_n\}$$

converges a.s., and since it is a deterministic sum, it converges! This last part of the argument was used in Section 12.4. $\square$

APPENDICES

Chapter A1
Appendix to Chapter 1

A1.1. A non-measurable subset A of S^1.

In the spirit of Banach and Tarski, although, of course, this relatively trivial example pre-dates theirs, we use the Axiom of Choice to show that

(a)
$$S^1 = \bigcup_{q \in \mathbf{Q}} A_q$$

where the A_q are disjoint sets, each of which may be obtained from any of the others by rotation. If the set $A = A_0$ has a 'length' then it is intuitively clear that result (a) would force

$$2\pi = \infty \times \text{length } (A),$$

an impossibility.

To construct the family $(A_q : q \in \mathbf{Q})$, proceed as follows. Regard S^1 as $\{e^{i\theta} : \theta \in \mathbf{R}\}$ inside $\mathbf{C}$. Define an equivalence relation $\sim$ on S^1 by writing $z \sim w$ if there exist α and β in $\mathbf{R}$ such that

$$z = e^{i\alpha}, \quad w = e^{i\beta}, \quad \alpha - \beta \in \mathbf{Q}.$$

Use the Axiom of Choice to produce a set A which has precisely one representative of each equivalence class. Define

$$A_q = e^{iq}A = \{e^{iq}z : z \in A\}.$$

Then the family $(A_q : q \in \mathbf{Q})$ has the desired properties. (Obviously, $\mathbf{Q}$ could be replaced by $\mathbf{Z}$ throughout the above argument.)

We do not bring this example to its fully rigorous conclusion. The remainder of this appendix is fully rigorous.

We now set out to prove Uniqueness Lemma 1.6.

A1.2. d-systems.

Let S be a set, and let $\mathcal{D}$ be a collection of subsets of S. Then $\mathcal{D}$ is called a *d-system* (*on S*) if

(a) $S \in \mathcal{D}$,

(b)   if $A, B \in \mathcal{D}$ and $A \subseteq B$ then $B \backslash A \in \mathcal{D}$,

(c)   if $A_n \in \mathcal{D}$ and $A_n \uparrow A$, then $A \in \mathcal{D}$.

Recall that $A_n \uparrow A$ means: $A_n \subseteq A_{n+1}(\forall n)$ and $\bigcup A_n = A$.

(d)   **Proposition.** *A collection Σ of subsets of S is a σ-algebra if and only if Σ is both a π-system and a d-system.*

Proof. The 'only if' part is trivial, so we prove only the 'if' part.

Suppose that Σ is both a π-system and a d-system, and that E, F and $E_n (n \in \mathbb{N}) \in \Sigma$. Then $E^c := S \backslash E \in \Sigma$, and

$$E \cup F = S \backslash (E^c \cap F^c) \in \Sigma.$$

Hence $G_n := E_1 \cup \ldots \cup E_n \in \Sigma$ and, since $G_n \uparrow \bigcup E_k$, we see that $\bigcup E_k \in \Sigma$. Finally,

$$\bigcap E_k = \left(\bigcup E_k^c \right)^c \in \Sigma. \qquad \square$$

Definition of $d(\mathcal{C})$. Suppose that $\mathcal{C}$ is a class of subsets of S. We define $d(\mathcal{C})$ to be the intersection of all d-systems which contain $\mathcal{C}$. Obviously, $d(\mathcal{C})$ is a d-system, the smallest d-system which contains $\mathcal{C}$. It is also obvious that

$$d(\mathcal{C}) \subseteq \sigma(\mathcal{C}).$$

A1.3. Dynkin's Lemma

▶▶   *If $\mathcal{I}$ is a π-system, then*

$$d(\mathcal{I}) = \sigma(\mathcal{I}).$$

Thus any d-system which contains a π-system contains the σ-algebra generated by that π-system.

Proof. Because of Proposition A1.2(d), we need only prove that $d(\mathcal{I})$ is a π-system.

Step 1: Let $\mathcal{D}_1 := \{B \in d(\mathcal{I}) : B \cap C \in d(\mathcal{I}), \forall C \in \mathcal{I}\}$. Because $\mathcal{I}$ is a π-system, $\mathcal{D}_1 \supseteq \mathcal{I}$. It is easily checked that $\mathcal{D}_1$ inherits the d-system structure from $d(\mathcal{I})$. [For, clearly, $S \in \mathcal{D}_1$. Next, if $B_1, B_2 \in \mathcal{D}_1$ and $B_1 \subseteq B_2$, then, for C in $\mathcal{I}$,

$$(B_2 \backslash B_1) \cap C = (B_2 \cap C) \backslash (B_1 \cap C);$$

and, since $B_2 \cap C \in d(\mathcal{I})$, $B_1 \cap C \in d(\mathcal{I})$ and $d(\mathcal{I})$ is a d-system, we see that $(B_2 \backslash B_1) \cap C \in d(\mathcal{I})$, so that $B_2 \backslash B_1 \in \mathcal{D}_1$. Finally, if $B_n \in \mathcal{D}_1 (n \in \mathbb{N})$ and $B_n \uparrow B$, then for $C \in \mathcal{I}$,

$$(B_n \cap C) \uparrow (B \cap C)$$

so that $B \cap C \in d(\mathcal{I})$ and $B \in \mathcal{D}_1$.] We have shown that $\mathcal{D}_1$ is a d-system which contains $\mathcal{I}$, so that (since $\mathcal{D}_1 \subseteq d(\mathcal{I})$ by its definition) $\mathcal{D}_1 = d(\mathcal{I})$.

Step 2: Let $\mathcal{D}_2 := \{A \in d(\mathcal{I}) : B \cap A \in d(\mathcal{I}), \forall B \in d(\mathcal{I})\}$. Step 1 showed that $\mathcal{D}_2$ contains $\mathcal{I}$. But, just as in Step 1, we can prove that $\mathcal{D}_2$ inherits the d-system structure from $d(\mathcal{I})$ and that therefore $\mathcal{D}_2 = d(\mathcal{I})$. But the fact that $\mathcal{D}_2 = d(\mathcal{I})$ says that $d(\mathcal{I})$ is a π-system. $\square$

A1.4. Proof of Uniqueness Lemma 1.6

Recall what the crucial Lemma 1.6 stated:

> Let S be a set. Let $\mathcal{I}$ be a π-system on S, and let $\Sigma := \sigma(\mathcal{I})$. Suppose that μ_1 and μ_2 are measures on (S, Σ) such that $\mu_1(S) = \mu_2(S) < \infty$ and $\mu_1 = \mu_2$ on $\mathcal{I}$. Then
>
> $$\mu_1 = \mu_2 \quad on \quad \Sigma.$$

Proof. Let

$$\mathcal{D} = \{F \in \Sigma : \mu_1(F) = \mu_2(F)\}.$$

Then $\mathcal{D}$ is a d-system on S. [Indeed, the fact that $S \in \mathcal{D}$ is given. If $A, B \in \mathcal{D}$, then

$$(*) \qquad \mu_1(B \backslash A) = \mu_1(B) - \mu_1(A) = \mu_2(B) - \mu_2(A) = \mu_2(B \backslash A),$$

so that $B \backslash A \in \mathcal{D}$. Finally, if $F_n \in \mathcal{D}$ and $F_n \uparrow F$, then by Lemma 1.10(a),

$$\mu_1(F) = \uparrow \lim \mu_1(F_n) = \uparrow \lim \mu_2(F_n) = \mu_2(F),$$

so that $F \in \mathcal{D}$.]

Since $\mathcal{D}$ is a d-system and $\mathcal{D} \supseteq \mathcal{I}$ by hypothesis, Dynkin's Lemma shows that $\mathcal{D} \supseteq \sigma(\mathcal{I}) = \Sigma$, and the result follows.   □

Notes. You should check that no circular argument is entailed by the use of Lemma 1.10 (this is *obvious*).

The reason for the insistence on finiteness in the condition $\mu_1(S) = \mu_2(S) < \infty$ is that we do not wish to try to claim at (*) that

$$\infty - \infty = \infty - \infty.$$

Indeed the Lemma 1.6 is false if '$< \infty$' is omitted – see Section A1.10 below.

We now aim to prove Carathéodory's Theorem 1.7.

A1.5. λ-sets: 'algebra' case

LEMMA

Let $\mathcal{G}_0$ be an algebra of subsets of S and let

$$\lambda : \mathcal{G}_0 \to [0, \infty]$$

with $\lambda(\emptyset) = 0$. Call an element L of $\mathcal{G}_0$ a λ-set if L 'splits every element of $\mathcal{G}_0$ properly':

$$\lambda(L \cap G) + \lambda(L^c \cap G) = \lambda(G), \quad \forall G \in \mathcal{G}_0.$$

Then the class $\mathcal{L}_0$ of λ-sets is an algebra, and λ is finitely additive on $\mathcal{L}_0$. Moreover, for disjoint $L_1, L_2, \ldots, L_n \in \mathcal{L}_0$ and G in $\mathcal{G}_0$,

$$\lambda\left(\bigcup_{k=1}^{n} (L_k \cap G) \right) = \sum_{k=1}^{n} \lambda(L_k \cap G).$$

Proof. Step 1: Let L_1 and L_2 be λ-sets, and let $L = L_1 \cap L_2$. We wish to prove that L is a λ-set.

Now $L^c \cap L_2 = L_2 \cap L_1^c$ and $L^c \cap L_2^c = L_2^c$. Hence, since L_2 is a λ-set, we have, for any G in $\mathcal{G}_0$,

$$\lambda(L^c \cap G) = \lambda(L_2 \cap L_1^c \cap G) + \lambda(L_2^c \cap G)$$

and, of course
$$\lambda(L_2^c \cap G) + \lambda(L_2 \cap G) = \lambda(G).$$

Since L_1 is a λ-set,

$$\lambda(L_2 \cap L_1^c \cap G) + \lambda(L \cap G) = \lambda(L_2 \cap G).$$

On adding the three equations just obtained, we see that

$$\lambda(L^c \cap G) + \lambda(L \cap G) = \lambda(G), \quad \forall G \in \mathcal{G}_0,$$

so that L is indeed a λ-set.

Step 2: Since, trivially, S is a λ-set, and the complement of a λ-set is a λ-set, it now follows that $\mathcal{L}_0$ is an algebra.

Step 3: If L_1 and L_2 are *disjoint* λ-sets and $G \in \mathcal{G}_0$, then

$$(L_1 \cup L_2) \cap L_1 = L_1, \quad (L_1 \cup L_2) \cap L_1^c = L_2,$$

so, since L_1 is a λ-set,

$$\lambda((L_1 \cup L_2) \cap G) = \lambda(L_1 \cap G) + \lambda(L_2 \cap G).$$

The proof is now easily completed. □

A1.6. Outer measures

Let $\mathcal{G}$ be a σ-algebra of subsets of S. A map

$$\lambda : \mathcal{G} \to [0, \infty]$$

is called an *outer measure* on $(S, \mathcal{G})$ if

(a) $\lambda(\emptyset) = 0$;

(b) λ is *increasing*: for $G_1, G_2 \in \mathcal{G}$ with $G_1 \subseteq G_2$,

$$\lambda(G_1) \leq \lambda(G_2);$$

(c) λ is *countably subadditive*: if (G_k) is any sequence of elements of $\mathcal{G}$, then

$$\lambda\left(\bigcup_k G_k\right) \leq \sum_k \lambda(G_k).$$

A1.7. Carathéodory's Lemma.

▶▶   *Let λ be an outer measure on the measurable space $(S,\mathcal{G})$. Then the λ-sets in $\mathcal{G}$ form a σ-algebra $\mathcal{L}$ on which λ is countably additive, so that $(S,\mathcal{L},\lambda)$ is a measure space.*

Proof. Because of Lemma A1.5, we need only show that if (L_k) is a disjoint sequence of sets in $\mathcal{L}$, then $L := \bigcup_k L_k \in \mathcal{L}$ and

(a) $$\lambda(L) = \sum_k \lambda(L_k).$$

By the subadditive property of λ, for $G \in \mathcal{G}$, we have

(b) $$\lambda(G) \leq \lambda(L \cap G) + \lambda(L^c \cap G).$$

Now let $M_n := \bigcup_{k \leq n} L_k$. Lemma A1.5 shows that $M_n \in \mathcal{L}$, so that

$$\lambda(G) = \lambda(M_n \cap G) + \lambda(M_n^c \cap G).$$

However, $M_n^c \supseteq L^c$, so that

(c) $$\lambda(G) \geq \lambda(M_n \cap G) + \lambda(L^c \cap G).$$

Lemma A1.5 now allows us to rewrite (c) as

$$\lambda(G) \geq \sum_{k \leq n} \lambda(L_k \cap G) + \lambda(L^c \cap G),$$

so that

(d) $$\lambda(G) \geq \sum_k \lambda(L_k \cap G) + \lambda(L^c \cap G)$$
$$\geq \lambda(L \cap G) + \lambda(L^c \cap G),$$

using the countably subadditive property of λ in the last step. On comparing (d) with (b), we see that equality must hold throughout (d) and (b), so that $L \in \mathcal{L}$; and then on taking $G = L$ we obtain result (a). $\square$

A1.8. Proof of Carathéodory's Theorem.

Recall that we need to prove the following.

Let S be a set, let Σ_0 be an algebra on S, and let

$$\Sigma := \sigma(\Sigma_0).$$

If μ_0 is a countably additive map $\mu_0 : \Sigma_0 \to [0, \infty]$, then there exists a measure μ on (S, Σ) such that

$$\mu = \mu_0 \ on \ \Sigma_0.$$

Proof. Step 1: Let $\mathcal{G}$ be the σ-algebra of *all* subsets of S. For $G \in \mathcal{G}$, define

$$\lambda(G) := \inf \sum_n \mu_0(F_n),$$

where the infimum is taken over all sequences (F_n) in Σ_0 with $G \subseteq \bigcup_n F_n$. We now prove that

(a) λ *is an outer measure on* $(S, \mathcal{G})$.

The facts that $\lambda(\emptyset) = 0$ and λ *is increasing* are obvious. Suppose that (G_n) is a sequence in $\mathcal{G}$, such that each $\lambda(G_n)$ is finite. Let $\varepsilon > 0$ be given. For each n, choose a sequence $(F_{n,k} : k \in \mathbf{N})$ of elements of Σ_0 such that

$$G_n \subseteq \bigcup_k F_{n,k}, \quad \sum_k \mu_0(F_{n,k}) < \lambda(G_n) + \varepsilon 2^{-n}.$$

Then $G := \bigcup G_n \subseteq \bigcup_n \bigcup_k F_{n,k}$, so that

$$\lambda(G) \leq \sum_n \sum_k \mu_0(F_{n,k}) < \sum_n \lambda(G_n) + \varepsilon.$$

Since ε is arbitrary, we have proved result (a).

Step 2: By Carathéodory's Lemma A1.7, λ is a measure on $(S, \mathcal{L})$, where $\mathcal{L}$ is the σ-algebra of λ-sets in $\mathcal{G}$. All we need show is that

(b) $\Sigma_0 \subseteq \mathcal{L}$, and $\lambda = \mu_0$ on Σ_0;

for then $\Sigma := \sigma(\Sigma_0) \subseteq \mathcal{L}$ and we can define μ to be the restriction of λ to (S, Σ).

Step 3: Proof that $\lambda = \mu_0$ on Σ_0.

Let $F \in \Sigma_0$. Then, clearly, $\lambda(F) \leq \mu_0(F)$. Now suppose that $F \subseteq \bigcup_n F_n$, where $F_n \in \Sigma_0$. As usual, we can define a sequence (E_n) of disjoint sets:

$$E_1 := F_1, \quad E_n = F_n \cap \left(\bigcup_{k<n} F_k \right)^c$$

such that $E_n \subseteq F_n$ and $\bigcup E_n = \bigcup F_n \supseteq F$. Then

$$\mu_0(F) = \mu_0 \left(\bigcup (F \cap E_n) \right) = \sum \mu_0(F \cap E_n),$$

by using the countable additivity of μ_0 on Σ_0. Hence

$$\mu_0(F) \leq \sum \mu_0(E_n) \leq \sum \mu_0(F_n),$$

so that $\lambda(F) \geq \mu_0(F)$. Step 3 is complete.

Step 4: Proof that $\Sigma_0 \subseteq \mathcal{L}$. Let $E \in \Sigma_0$ and $G \in \mathcal{G}$. Then there exists a sequence (F_n) in Σ_0 such that $G \subseteq \bigcup_n F_n$, and

$$\sum_n \mu_0(F_n) \leq \lambda(G) + \varepsilon.$$

Now, by definition of λ,

$$\sum_n \mu_0(F_n) = \sum_n \mu_0(E \cap F_n) + \sum_n \mu_0(E^c \cap F_n)$$
$$\geq \lambda(E \cap G) + \lambda(E^c \cap G),$$

since $E \cap G \subseteq \bigcup(E \cap F_n)$ and $E^c \cap G \subseteq \bigcup(E^c \cap F_n)$. Thus, since ε is arbitrary,

$$\lambda(G) \geq \lambda(E \cap G) + \lambda(E^c \cap G).$$

However, since λ is subadditive,

$$\lambda(G) \leq \lambda(E \cap G) + \lambda(E^c \cap G).$$

We see that E is indeed a λ-set. $\qquad\qquad\qquad\qquad\qquad\qquad\square$

A1.9. Proof of the existence of Lebesgue measure on $((0,1], \mathcal{B}(0,1])$.

Recall the set-up in Section 1.8. Let $S = (0,1]$. For $F \subseteq S$, say that $F \in \Sigma_0$ if F may be written as a finite union

$$(*) \qquad\qquad F = (a_1, b_1] \cup \ldots \cup (a_r, b_r]$$

where $r \in \mathbb{N}$, $0 \leq a_1 \leq b_1 \leq \ldots \leq a_r \leq b_r \leq 1$. Then (as you should convince yourself) Σ_0 is an algebra on $(0,1]$ and

$$\Sigma := \sigma(\Sigma_0) = \mathcal{B}(0,1].$$

(We write $\mathcal{B}(0,1]$ rather than $\mathcal{B}((0,1])$.) For F as at $(*)$, let

$$\mu_0(F) = \sum_{k \leq r} (b_k - a_k).$$

Of course, a set F may have different expressions as a finite disjoint union of the form $(*)$: for example,

$$(0,1] = (0, \tfrac{1}{2}] \cup (\tfrac{1}{2}, 1].$$

However, it is easily seen that μ_0 is well defined on Σ_0 and that μ_0 is finitely additive on Σ_0. While this is obvious from a picture, you might (or might not) wish to consider how to make the intuitive argument into a formal proof.

The key thing is to prove that μ_0 *is countably additive on* Σ_0. So, suppose that (F_n) is a sequence of disjoint elements of Σ_0 with union F in Σ_0. We know that if $G_n = \bigcup_{k=1}^{n} F_k$, then

$$\mu_0(G_n) = \sum_{k=1}^{n} \mu_0(F_k) \quad \text{and} \quad G_n \uparrow F.$$

To prove that μ_0 is countably additive it is enough to show that $\mu_0(G_n) \uparrow \mu_0(F)$, for then

$$\mu_0(F) =\uparrow \lim \mu_0(G_n) =\uparrow \lim_n \sum_{k=1}^{n} \mu_0(F_k) = \sum \mu_0(F_k).$$

Let $H_n = F \backslash G_n$. Then $H_n \in \Sigma_0$ and $H_n \downarrow \emptyset$. We need only prove that

$$\mu_0(H_n) \downarrow 0;$$

for then

$$\mu_0(G_n) = \mu_0(F) - \mu_0(H_n) \uparrow \mu_0(F).$$

It is clear that an alternative (and final!) rewording of what we need to show is the following:

(a)   *if (H_n) is a decreasing sequence of elements of Σ_0 such that for some $\varepsilon > 0$,*

$$\mu_0(H_n) \geq 2\varepsilon, \quad \forall n,$$

then $\bigcap_k H_k \neq \emptyset$.

Proof of (a). It is obvious from the definition of Σ_0 that, for each $k \in \mathbf{N}$, we can choose $J_k \in \Sigma_0$ such that, with $\bar{J}_k$ denoting the closure of J_k,

$$\bar{J}_k \subseteq H_k \quad \text{and} \quad \mu(H_k \backslash J_k) \leq \varepsilon 2^{-k}.$$

But then (recall that $H_n \downarrow$)

$$\mu_0\left(H_n \backslash \bigcap_{k \leq n} J_k\right) \leq \mu_0\left(\bigcup_{k \leq n}(H_k \backslash J_k)\right) \leq \sum_{k \leq n} \varepsilon 2^{-k} < \varepsilon.$$

Hence, since $\mu_0(H_n) > 2\varepsilon$, $\forall n$, we see that for every n,

$$\mu_0\left(\bigcap_{k \leq n} J_k\right) > \varepsilon,$$

and hence $\bigcap_{k \leq n} J_k$ is non-empty. *A fortiori* then, for every n,

$$K_n := \bigcap_{k \leq n} \bar{J}_k \quad \text{is non-empty.}$$

That

(b) $$\bigcap \bar{J}_k \neq \emptyset \quad \text{(whence } \bigcap H_k \neq \emptyset)$$

now follows from the Heine-Borel theorem: for if (b) is false, then $((\bar{J}_k)^c : k \in \mathbf{N})$ gives a covering of $[0, 1]$ by open sets with no finite subcovering. Alternatively, we can argue directly as follows. For each n, choose a point x_n in the non-empty set K_n. Since each x_n belongs to the compact set $\bar{J}_1$, we can find a subsequence (n_q) and a point x of $\bar{J}_1$ such that $x_{n_q} \to x$. However, for each k, $x_{n_q} \in \bar{J}_k$ for all but finitely many q, and since $\bar{J}_k$ is compact, it follows that $x \in \bar{J}_k$. Hence $x \in \bigcap_k \bar{J}_k$, and property (b) holds.

$\qquad\qquad\qquad\qquad\qquad\qquad\qquad\qquad\qquad\qquad\qquad\qquad\qquad$ $\square$

Since μ_0 is countably additive on Σ_0 and $\mu_0(0,1] < \infty$, it follows that μ_0 has a unique extension to a measure μ on $((0,1], \mathcal{B}(0,1])$. This is Lebesgue measure Leb on $((0,1], \mathcal{B}(0,1])$.

The μ_0-sets form a σ-algebra strictly larger than $\mathcal{B}(0,1]$, namely the σ-algebra of Lebesgue measurable subsets of $(0,1]$. See Section A1.11.

A1.10. Example of non-uniqueness of extension

With (S, Σ_0) as in Section A1.9, suppose that for $F \in \Sigma_0$,

(a) $\qquad\qquad\qquad\qquad \nu_0(F) := \begin{cases} 0 & \text{if } F = \emptyset, \\ \infty & \text{if } F \neq \emptyset. \end{cases}$

The Carathéodory extension of ν_0 will be obtained as the obvious extension of (a) to Σ. However, another extension $\tilde{\nu}$ is given by

$$\tilde{\nu}(F) = \text{number of elements in } F.$$

A1.11. Completion of a measure space

In fact (apart from an 'aside' on the Riemann integral), we do not need completions in this book.

Suppose that (S, Σ, μ) is a measure space. Define a class $\mathcal{N}$ of subsets of S as follows:

$N \in \mathcal{N}$ if and only if $\exists Z \in \Sigma$ such that $N \subseteq Z$ and $\mu(Z) = 0$.

It is sometimes philosophically satisfying to be able to make precise the idea that 'N in $\mathcal{N}$ is μ-measurable and $\mu(N) = 0$'. This is done as follows. For any subset F of S, write

$$F \in \Sigma^*$$

if $\exists E, G \in \Sigma$ such that $E \subseteq F \subseteq G$ and $\mu(G \setminus E) = 0$. It is very easy to show that Σ^* is a σ-algebra on S and indeed that $\Sigma^* = \sigma(\Sigma, \mathcal{N})$. With obvious notation we define for $F \in \Sigma^*$,

$$\mu^*(F) = \mu(E) = \mu(G),$$

it being easy to check that μ^* is well defined. Moreover, it is no problem to prove that (S, Σ^*, μ^*) is a measure space, the *completion* of (S, Σ, μ).

For parts of advanced probability, it is essential to complete the basic probability triple $(\Omega, \mathcal{F}, \mathbf{P})$. In other parts of probability, when (for example) S is topological, $\Sigma = \mathcal{B}(S)$, and we wish to consider several different measures on (S, Σ), it is meaningless to insist on completion.

If we begin with $([0,1], \mathcal{B}[0,1], \mathrm{Leb})$, then $\mathcal{B}[0,1]^*$ is the σ-algebra of what are called *Lebesgue-measurable* sets of [0,1]. Then, for example, a function $f : [0,1] \to [0,1]$ is Lebesgue-measurable if the inverse image of every *Borel* set is Lebesgue-measurable: it need not be true that the inverse image of a Lebesgue-measurable set is Lebesgue-measurable.

A1.12. The Baire category theorem

In Section 1.11, we studied a subset H of $S := [0,1]$ such that

(i) $H = \bigcap_k G_k$ for a sequence (G_k) of open subsets of S,

(ii) $H \supseteq V$, where $V = \mathbf{Q} \cap S$.

If H were countable: $H = \{h_r : r \in \mathbf{N}\}$, then we would have

(a) $$S = H \cup H^c = \left(\bigcup_r \{h_r\}\right) \cup \left(\bigcup G_k^c\right)$$

expressing S as a countable union

(b) $$S = \bigcup_n F_n$$

of closed sets where no F_n contains an open interval. [Since $V \subseteq G_k$ for every k, $G_k^c \subseteq V^c$ so that G_k^c contains only irrational points in S.]

However, the Baire category theorem states that

if a complete metric space S may be written as a union of a countable sequence of closed sets:

$$S = \bigcup F_n$$

then some F_n contains an open ball.

Thus the set H must be uncountable.

The Baire category theorem has fundamental applications in functional analysis, and some striking applications to probability too!

Proof of the Baire category theorem. Assume for the purposes of contradiction that no F_n contains an open ball. Since F_1^c is a non-empty open subset of S, we can find x_1 in S and $\varepsilon_1 > 0$ such that

$$B(x_1, \varepsilon_1) \subseteq F_1^c,$$

$B(x_1, \varepsilon_1)$ denoting the open ball of radius ε_1 centred at x_1. Now F_2 contains no open ball, so that the open set

$$U_2 := B(x_1, 2^{-1}\varepsilon_1) \cap F_2^c$$

is non-empty, and we can find x_2 in U_2 and $\varepsilon_2 > 0$ such that

$$B(x_2, \varepsilon_2) \subseteq U_2, \quad \varepsilon_2 < 2^{-1}\varepsilon_1.$$

Inductively, choose a sequence (x_n) in S and (ε_n) in $(0, \infty)$ so that we have $\varepsilon_{n+1} < 2^{-1}\varepsilon_n$ and

$$B(x_{n+1}, \varepsilon_{n+1}) \subseteq U_{n+1} := B(x_n, 2^{-1}\varepsilon_n) \cap F_{n+1}^c.$$

Since $d(x_n, x_{n+1}) < 2^{-1}\varepsilon_n$, it is obvious from the triangle law that (x_n) is Cauchy, so that $x := \lim x_n$ exists, and that

$$x \in \bigcap B(x_n, \varepsilon_n) \subseteq \bigcap F_n^c$$

contradicting the fact that $\bigcup F_n = S$. $\square$

Chapter A3
Appendix to Chapter 3

A3.1. Proof of the Monotone-Class Theorem 3.14

Recall the statement of the theorem.

▶▶   *Let $\mathcal{H}$ be a class of bounded functions from a set S into $\mathbb{R}$ satisfying the following conditions:*

     (i)   *$\mathcal{H}$ is a vector space over $\mathbb{R}$;*

     (ii)   *the constant function 1 is an element of $\mathcal{H}$;*

     (iii)   *if (f_n) is a sequence of non-negative functions in $\mathcal{H}$ such that $f_n \uparrow f$ where f is a bounded function on S, then $f \in \mathcal{H}$.*

Then if $\mathcal{H}$ contains the indicator function of every set in some π-system $\mathcal{I}$, then $\mathcal{H}$ contains every bounded $\sigma(\mathcal{I})$-measurable function on S.

Proof. Let $\mathcal{D}$ be the class of sets F in S such that $I_F \in \mathcal{H}$. It is immediate from (i) - (iii) that $\mathcal{D}$ is a d-system. Since $\mathcal{D}$ contains the π-system $\mathcal{I}$, $\mathcal{D}$ contains $\sigma(\mathcal{I})$.

     Suppose that f is a $\sigma(\mathcal{I})$-measurable function such that for some K in $\mathbb{N}$,

$$0 \le f(s) \le K, \quad \forall s \in S.$$

For $n \in \mathbb{N}$, define

$$f_n(s) := \sum_{i=0}^{K2^n} i2^{-n} I_{A(n,i)}(s),$$

where

$$A(n,i) := \{s : i2^{-n} \le f(s) < (i+1)2^{-n}\}.$$

Since f is $\sigma(\mathcal{I})$-measurable, every $A(n,i) \in \sigma(\mathcal{I})$, so that $I_{A(n,i)} \in \mathcal{H}$. Since $\mathcal{H}$ is a vector space, every $f_n \in \mathcal{H}$. But $0 \le f_n \uparrow f$, so that $f \in \mathcal{H}$.

If $f \in b\sigma(\mathcal{I})$, we may write $f = f^+ - f^-$, where $f = \max(f, 0)$ and $f^- = \max(-f, 0)$. Then $f^+, f^- \in b\sigma(\mathcal{I})$ and $f^+, f^- \geq 0$, so that $f^+, f^- \in \mathcal{H}$ by what we established above.   □

A3.2. Discussion of generated σ-algebras

This is one of those situations in which it is actually easier to understand things in a more formal abstract setting. So, suppose that

Ω and S are sets, and that $Y : \Omega \to S$;

Σ is a σ-algebra on S:

$X : \Omega \to \mathbb{R}$.

Because Y^{-1} preserves all set operations,

$$Y^{-1}\Sigma := \{Y^{-1}B : B \in \Sigma\}$$

is a σ-algebra on Ω, and because it is tautologically the smallest σ-algebra $\mathcal{Y}$ on Ω such that Y is $\mathcal{Y}/\Sigma$ measurable (in that $Y^{-1} : \Sigma \to \mathcal{Y}$), we call it $\sigma(Y)$:

$$\sigma(Y) = Y^{-1}\Sigma.$$

LEMMA

(a) X is $\sigma(Y)$-measurable if and only if

$$X = f(Y)$$

   where f is a Σ-measurable function from S to $\mathbb{R}$.

Note. The 'if' part is just the Composition Lemma.

Proof of 'only if' part. It is enough to prove that

(b) $X \in b\sigma(Y)$ if and only if $\exists f \in b\Sigma$ such that $X = f(Y)$.

(Otherwise, consider arc tan X, for example.)

Though we certainly do not need the Monotone-Class Theorem to prove (b), we may as well use it.

So define $\mathcal{H}$ to be the class of all bounded functions X on Ω such that $X = f(Y)$ for some $f \in b\Sigma$. Taking $\mathcal{I} = \sigma(Y)$, note that if $F \in \mathcal{I}$ then $F = Y^{-1}B$ for some B in Σ, so that

$$I_F(\omega) = I_B(Y(\omega)),$$

so that $I_F \in \mathcal{H}$. That $\mathcal{H}$ is a vector space containing constants is obvious.

Finally, suppose that (X_n) is a sequence of elements of $\mathcal{H}$ such that, for some positive real constant K,

$$0 \le X_n \uparrow X \le K.$$

For each n, $X_n = f_n(Y)$ for some f_n in $b\Sigma$. Define $f := \limsup f_n$, so that $f \in b\Sigma$. Then $X = f(Y)$.   □

One has to be very careful about what Lemma (a) *means in practice.*

To be sure, result (3.13,b) is the special case when $(S, \Sigma) = (\mathbf{R}, \mathcal{B})$.

Discussion of (3.13,c). Suppose that $Y_k : \Omega \to \mathbf{R}$ for $1 \le k \le n$. We may define a map $Y : \Omega \to \mathbf{R}^n$ via

$$Y(\omega) := (Y_1(\omega), \ldots, Y_n(\omega)) \in \mathbf{R}^n.$$

The problem mentioned at (3.13,d) and in the Warning following it shows up here because, before we can apply Lemma (a) to prove (3.13,c), we need to prove that

$$\sigma(Y_1, \ldots, Y_n) := \sigma(Y_k^{-1}\mathcal{B}(\mathbf{R}) : 1 \le k \le n) = Y^{-1}\mathcal{B}(\mathbf{R}^n) =: \sigma(Y).$$

[This amounts to proving that the product σ-algebra $\prod_{1 \le k \le n} \mathcal{B}(\mathbf{R})$ is the same as $\mathcal{B}(\mathbf{R}^n)$. See Section 8.5.] Now $Y_k = \gamma_k \circ Y$, where γ_k is the (continuous, hence Borel) 'k^{th} coordinate' map on $\mathbf{R}^n$, so that Y_k is $\sigma(Y)$-measurable. On the other hand, every open subset of $\mathbf{R}^n$ is a countable union of open rectangles $G_1 \times \cdots \times G_n$ where each G_k is a subinterval of $\mathbf{R}$, and since

$$\{Y \in G_1 \times \cdots \times G_n\} = \bigcap \{Y_k \in G_k\} \in \sigma(Y_1, \ldots, Y_n),$$

things do work out.   □

You can already see why we are in an appendix, and why we skip discussion of (3.13,d).

Appendix to Chapter 4

This appendix gives the statement of Strassen's Law of the Iterated Logarithm. Section A4.3 treats the completely different topic of constructing a rigorous model for a Markov chain.

A4.1. Kolmogorov's Law of the Iterated Logarithm

THEOREM

Let $X_1, X_2, \ldots$ be IID RVs each with mean 0 and variance 1.

Let $S_n := X_1 + X_2 + \cdots + X_n$. Then, almost surely,

$$\limsup \frac{S_n}{\sqrt{2n \log \log n}} = +1, \quad \liminf \frac{S_n}{\sqrt{2n \log \log n}} = -1.$$

This result already gives very precise behaviour on the big values of partial sums. See Section 14.7 for proof in the case when the X's are normally distributed.

A4.2. Strassen's Law of the Iterated Logarithm

Strassen's Law is a staggering extension of Kolmogorov's result.

Let (X_n) and (S_n) be as in the previous section. For each ω, let the map $t \mapsto S_t(\omega)$ on $[0, \infty)$ be the linear interpolation of the map $n \mapsto S_n(\omega)$ on $\mathbf{Z}^+$, so that

$$S_t(\omega) := (t - n)S_{n+1}(\omega) + (n + 1 - t)S_n(\omega), \qquad t \in [n, n+1).$$

With Kolmogorov's result in mind, define

$$Z_n(t, \omega) := \frac{S_{nt}(\omega)}{\sqrt{2n \log \log n}}, \qquad t \in [0, 1],$$

so that $t \mapsto Z_n(t, \omega)$ on $[0,1]$ is a rescaled version of the random walk S run up to time n. Say that a function $t \mapsto f(t, \omega)$ is in the set $K(\omega)$ of limiting shapes of the path associated with ω if there is a sequence $n_1(\omega), n_2(\omega), \ldots$ in $\mathbf{N}$ such that

$$Z_n(t, \omega) \to f(t, \omega) \text{ uniformly in } t \in [0, 1].$$

Now let K consist of those functions f in $C[0, 1]$ which can be written in the Lebesgue-integral form

$$f(t) = \int_0^t h(s)ds \quad \text{where} \quad \int_0^1 h(s)^2 ds \leq 1.$$

Strassen's Theorem
$$\mathbf{P}[K(\omega) = K] = 1.$$

Thus, (almost) all paths have the same limiting shapes. Khinchine's law follows from Strassen's precisely because (Exercise!)

$$\sup\{f(1): f \in K\} = 1, \quad \inf\{f(1): f \in K\} = -1.$$

However, the only element of K for which $f(1) = 1$ is the function $f(t) = t$, so the big values of S occur when the whole path (when rescaled) looks like a line of slope 1.

Almost every path will, in its Z rescaling, look infinitely often like the function t and infinitely often like the function $-t$; etc., etc.

References. For a highly-motivated classical proof of Strassen's Law, see Freedman (1971). For a proof for Brownian motion based on the powerful theory of large deviations, see Stroock (1984).

A4.3. A model for a Markov chain

Let E be a countable set; let μ be a probability measure on $(E, \mathcal{E})$, where $\mathcal{E}$ denotes the set of all subsets of E; and let P denote a stochastic $E \times E$ matrix as in Section 4.8.

Complicating the notation somewhat for reasons we shall discover later, we wish to construct a probability triple $(\tilde{\Omega}, \tilde{\mathcal{F}}, \tilde{\mathbf{P}}^\mu)$ carrying an E-valued stochastic process $(\tilde{Z}_n: n \in \mathbf{Z}^+)$ such that for $n \in \mathbf{Z}^+$ and $i_0, i_1, \ldots, i_n \in E$, we have

$$\tilde{\mathbf{P}}^\mu(\tilde{Z}_0 = i_0; \ldots; \tilde{Z}_n = i_n) = \mu_{i_0} p_{i_0 i_1} \cdots p_{i_{n-1} i_n}.$$

The trick is to make $(\tilde{\Omega}, \tilde{\mathcal{F}}, \tilde{\mathbf{P}}^\mu)$ carry *independent* E-valued variables

$$(\tilde{Z}_0; \tilde{Y}(i,n): i \in E; n \in \mathbf{N})$$

$\tilde{Z}_0$ having law μ and such that

$$\tilde{\mathbf{P}}^\mu(\tilde{Y}(i,n) = j) = p(i,j), \quad (i,j \in \mathbf{E}).$$

We can obviously do this via the construction in Section 4.6.

For $\tilde{\omega} \in \tilde{\Omega}$ and $n \in \mathbf{N}$, define

$$\tilde{Z}_n(\tilde{\omega}) := \tilde{Y}(\tilde{Z}_{n-1}(\tilde{\omega}), n);$$

and that's it!

Chapter A5
Appendix to Chapter 5

Our task is to prove the Monotone-Convergence Theorem 5.3. We need an elementary preliminary result.

A5.1. Doubly monotone arrays

Proposition. *Let*

$$(y_n^{(r)} : r \in \mathbf{N}, n \in \mathbf{N})$$

be an array of numbers in $[0, \infty]$ which is doubly monotone:

for fixed r, $y_n^{(r)} \uparrow$ as $n \uparrow$ so that $y^{(r)} := \uparrow \lim_n y_n^{(r)}$ exists;

for fixed n, $y_n^{(r)} \uparrow$ as $r \uparrow$ so that $y_n := \uparrow \lim_r y_n^{(r)}$ exists.

Then

$$y^{(\infty)} := \uparrow \lim_r y^{(r)} = \uparrow \lim_n y_n =: y_\infty.$$

Proof. The result is almost trivial. By replacing each $(y_n^{(r)})$ by arc tan $y_n^{(r)}$, we can assume that the $y^{(\cdot)}$ are uniformly bounded.

Let $\varepsilon > 0$ be given. Choose n_0 such that $y_{n_0} > y_\infty - \frac{1}{2}\varepsilon$. Then choose r_0 such that $y_{n_0}^{(r_0)} > y_{n_0} - \frac{1}{2}\varepsilon$. Then

$$y^{(\infty)} \geq y^{(r_0)} \geq y_{n_0}^{(r_0)} > y_\infty - \varepsilon,$$

so that $y^{(\infty)} \geq y_\infty$. Similarly, $y_\infty \geq y^{(\infty)}$. $\qquad \square$

A5.2. The key use of Lemma 1.10(a)

This is where the fundamental monotonicity property of measures is used. Please re-read Section 5.1 at this stage.

LEMMA

(a)   *Suppose that $A \in \Sigma$ and that*

$$h_n \in SF^+ \quad and \quad h_n \uparrow I_A.$$

Then $\mu_0(h_n) \uparrow \mu(A)$.

Proof. From (5.1,e), $\mu_0(h_n) \leq \mu(A)$, so we need only prove that

$$\liminf \mu_0(h_n) \geq \mu(A).$$

Let $\varepsilon > 0$, and define $A_n := \{s \in A : h_n(s) > 1 - \varepsilon\}$. Then $A_n \uparrow A$ so that, by Lemma 1.10(a), $\mu(A_n) \uparrow \mu(A)$. But

$$(1 - \varepsilon)I_{A_n} \leq h_n$$

so that, by (5.1,e), $(1 - \varepsilon)\mu(A_n) \leq \mu_0(h_n)$. Hence

$$\liminf \mu_0(h_n) \geq (1 - \varepsilon)\mu(A).$$

Since this is true for every $\varepsilon > 0$, the result follows.   □

LEMMA

(b)   *Suppose that $f \in SF^+$ and that*

$$g_n \in SF^+ \quad and \quad g_n \uparrow f.$$

Then $\mu_0(g_n) \uparrow \mu_0(f)$.

Proof. We can write f as a finite sum $f = \sum a_k I_{A_k}$ where the sets A_k are disjoint and each $a_k > 0$. Then

$$a_k^{-1} I_{A_k} g_n \uparrow I_{A_k} \qquad (n \uparrow \infty),$$

and the result follows from Lemma (a).   □

A5.3. 'Uniqueness of integral'

LEMMA

(a)   *Suppose that $f \in (m\Sigma)^+$ and that we have two sequences $(f^{(r)})$ and (f_n) of elements of SF^+ such that*

$$f^{(r)} \uparrow f, \quad f_n \uparrow f.$$

Then

$$\uparrow \lim \mu_0(f^{(r)}) = \uparrow \lim \mu_0(f_n).$$

Proof. Let $f_n^{(r)} := f^{(r)} \wedge f_n$. Then as $r \uparrow \infty$, $f_n^{(r)} \uparrow f_n$, and as $n \uparrow \infty$, $f_n^{(r)} \uparrow f^{(r)}$. Hence, by Lemma A5.2(b),

$$\mu_0(f_n^{(r)}) \uparrow \mu_0(f_n) \text{ as } r \uparrow \infty,$$
$$\mu_0(f_n^{(r)}) \uparrow \mu_0(f^{(r)}) \text{ as } n \uparrow \infty.$$

The result now follows from Proposition A5.1.   □

Recall from Section 5.2 that for $f \in (m\Sigma)^+$, we define

$$\mu(f) := \sup\{\mu_0(h) : h \in SF^+; h \leq f\} \leq \infty.$$

By definition of $\mu(f)$, we may choose a sequence h_n in SF^+ such that $h_n \leq f$ and $\mu_0(h_n) \uparrow \mu(f)$. Let us also choose a sequence (g_n) of elements of SF^+ such that $g_n \uparrow f$. (We can do this via the 'staircase function' in Section 5.3.) Now let

$$f_n := \max(g_n, h_1, h_2, \dots, h_n).$$

Then $f_n \in SF^+$, $f_n \leq f$, and since $f_n \geq g_n$, $f_n \uparrow f$. Since $f_n \leq f$, $\mu_0(f_n) \leq \mu(f)$, and since $f_n \geq h_n$, we see that

$$\mu_0(f_n) \uparrow \mu(f).$$

On combining this fact with Lemma (a), we obtain the next result. (Lemma (a) 'changes our particular sequence to *any* sequence'.)

LEMMA

(b)   *Let $f \in (m\Sigma)^+$ and let (f_n) be any sequence in SF^+ such that $f_n \uparrow f$. Then*

$$\mu(f_n) = \mu_0(f_n) \uparrow \mu(f).$$

A5.4. Proof of the Monotone-Convergence Theorem

Recall the statement:

   Let (f_n) be a sequence of elements of $(m\Sigma)^+$ such that $f_n \uparrow f$. Then

$$\mu(f_n) \uparrow \mu(f).$$

Proof. Let $\alpha^{(r)}$ denote the rth staircase function defined in Section 5.3. Now set $f_n^{(r)} := \alpha^{(r)}(f_n)$, $f^{(r)} := \alpha^{(r)}(f)$. Since $\alpha^{(r)}$ is left-continuous, $f_n^{(r)} \uparrow f^{(r)}$ as $n \uparrow \infty$. Since $\alpha^{(r)}(x) \uparrow x$, $\forall x$, $f_n^{(r)} \uparrow f_n$ as $r \uparrow \infty$. By Lemma A5.2(b), $\mu(f_n^{(r)}) \uparrow \mu(f^{(r)})$ as $n \uparrow \infty$; and by Lemma A5.3(b), $\mu(f_n^{(r)}) \uparrow \mu(f_n)$ as $r \uparrow \infty$. We also know from Lemma A5.3(b) that $\mu(f^{(r)}) \uparrow \mu(f)$. The result now follows from Proposition A5.1.   □

Chapter A9
Appendix to Chapter 9

This chapter is solely devoted to the proof of the 'infinite-product' Theorem 8.7. It may be read after Section 9.10. It is probably something which a keen student who has read all previous appendices should study with a tutor.

A9.1. Infinite products: setting things up

Let $(\Lambda_n :\in \mathbb{N})$ be a sequence of probability measures on $(\mathbb{R}, \mathcal{B})$. Let

$$\Omega := \prod_{n \in \mathbb{N}} \mathbb{R},$$

so that a typical element ω of Ω is a sequence $\omega = (\omega_n : n \in \mathbb{N})$ of elements of $\mathbb{R}$. Define $X_n(\omega) := \omega_n$, and set

$$\mathcal{F}_n := \sigma(X_1, X_2, \ldots, X_n).$$

The typical element F_n of $\mathcal{F}_n$ has the form

(a) $$F_n = G_n \times \prod_{k > n} \mathbb{R}, \qquad G_n \in \prod_{1 \leq k \leq n} \mathcal{B}.$$

Fubini's Theorem shows that on the *algebra* (NOT σ-algebra)

$$\mathcal{F}^- = \bigcup \mathcal{F}_n,$$

we may unambiguously use (a) to define a map $\mathbb{P}^- : \mathcal{F}^- \to [0, 1]$ via

(b) $$\mathbb{P}^-(F_n) = (\Lambda_1 \times \cdots \times \Lambda_n)(G_n),$$

and that $\mathbb{P}^-$ is *finitely additive on the algebra* $\mathcal{F}^-$.

However, for each fixed n,

(c) $(\Omega, \mathcal{F}_n, \mathbf{P}^-)$ *is a bona fide probability triple which may be identified with* $\prod_{1 \le k \le n}(\mathbf{R}, \mathcal{B}, \Lambda_k)$ *via* (a) *and* (b). *Moreover,* $X_1, X_2, \ldots, X_n$ *are independent RVs on* $(\Omega, \mathcal{F}_n, \mathbf{P}^-)$.

We want to prove that

(d) $\mathbf{P}^-$ *is countably additive on* $\mathcal{F}^-$

(obviously with the intention of using Carathéodory's Theorem 1.7). Now we know from our proof of the existence of Lebesgue measure (see (A1.9,a)) that it is enough to show that

(e) *if* (H_r) *is a sequence of sets in* $\mathcal{F}^-$ *such that* $H_r \supseteq H_{r+1}, \forall r$, *and if for some* $\varepsilon > 0$, $\mathbf{P}^-(H_r) \ge \varepsilon$ *for every* r, *then* $\bigcap H_r \ne \emptyset$.

A9.2. Proof of (A9.1,e)

Step 1: For every r, there is some $n(r)$ such that $H_r \in \mathcal{F}_{n(r)}$ and so

$$I_{H_r}(\omega) = h_r(\omega_1, \omega_2, \ldots, \omega_{n(r)}) \quad \text{for some} \quad h_r \in b\mathcal{B}^{n(r)}.$$

Recall that $X_k(\omega) = \omega_k$, and look again at Section A3.2.

Step 2: We have

(a0) $$\mathbf{E}^- h_r(X_1, X_2, \ldots, X_{n(r)}) \ge \varepsilon, \forall r,$$

because the left-hand side of (a0) is exactly $\mathbf{P}^-(H_r)$. If we work within the probability triple $(\Omega, \mathcal{F}_{n(r)}, \mathbf{P}^-)$, then we know from Section 9.10 that

$$\gamma_r(\omega) := g_r(\omega_1) := \mathbf{E}^- h_r(\omega_1, X_2, X_3, \ldots, X_{n(r)})$$

is an explicit version of the conditional expectation of I_{H_r} given $\mathcal{F}_1$, and

$$\varepsilon \le \mathbf{P}^-(H_r) = \mathbf{E}^-(\gamma_r) = \Lambda_1(g_r).$$

Now, $0 \le g_r \le 1$, so that

$$\varepsilon \le \Lambda_1(g_r) \le 1\Lambda_1\{g_r \ge \varepsilon 2^{-1}\} + \varepsilon 2^{-1}\Lambda_1\{g_r^{(1)} \le \varepsilon 2^{-1}\}$$
$$\le \Lambda_1\{g_r \ge \varepsilon 2^{-1}\} + \varepsilon 2^{-1}.$$

Thus

$$\Lambda_1\{g_r \ge \varepsilon 2^{-1}\} \ge 2^{-1}\varepsilon.$$

Step 3: However, since $H_r \supseteq H_{r+1}$, we have (working within $(\Omega, \mathcal{F}_m, \mathbf{P}^-)$ where both H_r and H_{r+1} are in $\mathcal{F}_m$)

$$g_r(\omega_1) \ge g_{r+1}(\omega_1), \quad \text{for every } \omega_1 \text{ in } \mathbf{R}.$$

Working on $(\mathbf{R}, \mathcal{B}, \Lambda_1)$ we have

$$\Lambda_1\{g_r \geq \varepsilon 2^{-1}\} \geq \varepsilon 2^{-1}, \forall r,$$

and

$$g_r \downarrow \text{ so that } \{g_r \geq \varepsilon 2^{-1}\} \downarrow;$$

and by Lemma 1.10(b) on the continuity from above of measures, we have

$$\Lambda_1\{\omega_1 : g_r(\omega_1) \geq \varepsilon 2^{-1}, \forall r\} \geq \varepsilon 2^{-1}.$$

Hence, there exists ω_1^* (say) in $\mathbf{R}$ such that

(a1) $$\mathbf{E}^- h_r(\omega_1^*, X_2, \ldots, X_{n(r)}) \geq \varepsilon 2^{-1}, \forall r.$$

Step 4: We now repeat Steps 2 and 3 applied to the situation in which

$$(X_1, X_2, \ldots) \quad \text{is replaced by} \quad (X_2, X_3, \ldots),$$

h_r is replaced by $h_r(\omega_1^*)$, where

$$(h_r(\omega_1^*))(\omega_2, \omega_3, \ldots) := h_r(\omega_1^*, \omega_2, \omega_3, \ldots).$$

We find that there exists ω_2^* in $\mathbf{R}$ such that

(a2) $$\mathbf{E}^- h_r(\omega_1^*, \omega_2^*, X_3, \ldots, X_{n(r)}) \geq \varepsilon 2^{-2}, \quad \forall r.$$

Proceeding inductively, we obtain a sequence

$$\omega^* = (\omega_n^* : n \in \mathbf{N})$$

with the property that

$$\mathbf{E}^- h_r(\omega_1^*, \omega_2^*, \ldots, \omega_{n(r)}^*) \geq \varepsilon 2^{-n(r)}, \forall r.$$

However,

$$h_r(\omega_1^*, \omega_2^*, \ldots, \omega_{n(r)}^*) = I_{H_r}(\omega^*),$$

and can only be 0 or 1. The only conclusion is that $\omega^* \in H_r, \forall r$; and it was exactly the existence of such an ω^* which we had to prove. $\square$

Appendix to Chapter 13

This chapter is devoted to comments on **modes of convergence**, regarded by many as good for the souls of students, and certainly easy to set examination questions on.

A13.1. Modes of convergence: definitions

Let $(X_n : n \in \mathsf{N})$ be a sequence of RVs and let X be a RV, all carried by our triple $(\Omega, \mathcal{F}, \mathbf{P})$. Let us collect together definitions known to us.

Almost sure convergence

We say that $X_n \to X$ **almost surely** if

$$\mathbf{P}(X_n \to X) = 1.$$

Convergence in probability

We say that $X_n \to X$ **in probability** if, for every $\varepsilon > 0$,

$$\mathbf{P}(|X_n - X| > \varepsilon) \to 0 \quad \text{as} \quad n \to \infty.$$

$\mathcal{L}^p$ convergence $(p \geq 1)$

We say that $X_n \to X$ in $\mathcal{L}^p$ if each X_n is in $\mathcal{L}^p$ and $X \in \mathcal{L}^p$ and

$$\|X_n - X\|_p \to 0 \quad \text{as} \quad n \to \infty,$$

equivalently,

$$\mathbf{E}(|X_n - X|^p) \to 0 \quad \text{as} \quad n \to \infty.$$

A13.2. Modes of convergence: relationships

Let me state the *facts*.

Convergence in probability is the weakest of the above forms of convergence. Thus

(a) $(X_n \to X,$ a.s.$) \Rightarrow (X_n \to X$ in prob$)$

(b) for $p \geq 1$,

$$(X_n \to X \text{ in } \mathcal{L}^p) \Rightarrow (X_n \to X \text{ in prob}).$$

No other implication between any two of our three forms of convergence is valid. But, of course, for $r \geq p \geq 1$,

(c) $(X_n \to X$ in $\mathcal{L}^r) \Rightarrow (X_n \to X$ in $\mathcal{L}^p)$.

If we know that 'convergence in probability is happening quickly' in that

(d) $\sum_n \mathbf{P}(|X_n - X| > \varepsilon) < \infty, \quad \forall \varepsilon > 0,$

then (BC1) allows us to conclude that $X_n \to X$, a.s.

The fact that property (d) implies a.s. convergence is used in proving the following result:

(e) *$X_n \to X$ in probability if and only if every subsequence of (X_n) contains a further subsequence along which we have almost sure convergence to X.*

The only other *useful* result is that

(f) *for $p \geq 1$, $X_n \to X$ in $\mathcal{L}^p$ if and only if the following two statements hold:*
  (i) *$X_n \to X$ in probability,*
  (ii) *the family $(|X_n|^p : n \geq 1)$ is UI.*

There is only one way to gain an understanding of the above facts, and that is to prove them yourself. The exercises under EA13 provide guidance if you need it.

Appendix to Chapter 14

We work with a filtered space $(\Omega, \mathcal{F}, \{\mathcal{F}_n : n \in \mathbf{Z}^+\}, \mathbf{P})$.

This chapter introduces the σ-algebra $\mathcal{F}_T$, where T is a stopping time. The idea is that $\mathcal{F}_T$ represents the information available to our observer at (or, if you prefer, immediately after) time T. The **Optional-Sampling Theorem** says that if X is a *uniformly integrable* supermartingale and S and T are stopping times with $S \leq T$, then we have the natural extension of the supermartingale property:

$$\mathsf{E}(X_T | \mathcal{F}_S) \leq X_S, \quad \text{a.s.}$$

A14.1. The σ-algebra $\mathcal{F}_T$, T a stopping time

Recall that a map $T : \Omega \to \mathbf{Z}^+ \cup \{\infty\}$ is called a *stopping time* if

$$\{T \leq n\} \in \mathcal{F}_n, \quad n \in \mathbf{Z}^+ \cup \{\infty\},$$

equivalently if

$$\{T = n\} \in \mathcal{F}_n, \quad n \in \mathbf{Z}^+ \cup \{\infty\}.$$

In each of the above statements, the '$n = \infty$' case follows automatically from the validity of the result for every n in $\mathbf{Z}^+$.

Let T be a stopping time. Then, for $F \subseteq \Omega$, we say that $F \in \mathcal{F}_T$ if

$$F \cap \{T \leq n\} \in \mathcal{F}_n, \quad n \in \mathbf{Z}^+ \cup \{\infty\},$$

equivalently if

$$F \cap \{T = n\} \in \mathcal{F}_n, \quad n \in \mathbf{Z}^+ \cup \{\infty\}.$$

Then $\mathcal{F}_T = \mathcal{F}_n$ if $T \equiv n$; $\mathcal{F}_T = \mathcal{F}_\infty$ if $T \equiv \infty$; and $\mathcal{F}_T \subseteq \mathcal{F}_\infty$ for every T.

You can easily check that $\mathcal{F}_T$ is a σ-algebra. You can also check that if S is another stopping time, then

$$\mathcal{F}_{S \wedge T} \subseteq \mathcal{F}_T \subseteq \mathcal{F}_{S \vee T}.$$

Hint. If $F \in \mathcal{F}_{S \wedge T}$, then

$$F \cap \{T = n\} = \bigcup_{k \leq n} F \cap \{S \wedge T = k\}. \qquad \square$$

Another detail that needs to be checked is that if X is an adapted process and T is a stopping time, then $X_T \in m\mathcal{F}_T$. Here, X_∞ is assumed defined in some way such that X_∞ is $\mathcal{F}_\infty$ measurable.

Proof. For $B \in \mathcal{B}$,

$$\{X_T \in B\} \cap \{T = n\} = \{X_n \in B\} \cap \{T = n\} \in \mathcal{F}_n. \qquad \square$$

A14.2. A special case of OST

LEMMA

> *Let X be a supermartingale. Let T be a stopping time such that, for some N in $\mathbf{N}$, $T(\omega) \leq N$, $\forall \omega$. Then $X_T \in \mathcal{L}^1(\Omega, \mathcal{F}_T, \mathbf{P})$ and*

$$\mathbf{E}(X_N | \mathcal{F}_T) \leq X_T.$$

Proof. Let $F \in \mathcal{F}_T$. Then

$$\mathbf{E}(X_N; F) = \sum_{n \leq N} \mathbf{E}(X_N; F \cap \{T = n\})$$

$$\leq \sum_{n \leq N} \mathbf{E}(X_n; F \cap \{T = n\}) = \mathbf{E}(X_T; F).$$

(Of course, the fact that $|X_T| \leq |X_1| + \cdots + |X_N|$ guarantees the result that $\mathbf{E}(|X_T|) < \infty$.) $\qquad \square$

A14.3. Doob's Optional-Sampling Theorem for UI martingales

▶▶   *Let M be a UI martingale. Then, for any stopping time T,*

$$\mathbf{E}(M_\infty | \mathcal{F}_T) = M_T, \quad \text{a.s.}$$

Corollary 1 (a new Optional-Stopping Theorem!)

If M is a UI martingale, and T is a stopping time, then $\mathsf{E}(|M_T|) < \infty$ and $\mathsf{E}(M_T) = \mathsf{E}(M_0)$.

Corollary 2

If M is a UI martingale and S and T are stopping times with $S \leq T$, then
$$\mathsf{E}(M_T | \mathcal{F}_S) = M_S, \quad \text{a.s.}$$

Proof of theorem. By Theorem 14.1 and Lemma A14.2, we have, for $k \in \mathbf{N}$,
$$\mathsf{E}(M_\infty | \mathcal{F}_k) = M_k, \quad \text{a.s.}, \quad \mathsf{E}(M_k | \mathcal{F}_{T \wedge k}) = M_{T \wedge k}, \quad \text{a.s.}$$

Hence, by the Tower Property,

$(*)$ $\mathsf{E}(M_\infty | \mathcal{F}_{T \wedge k}) = M_{T \wedge k}, \quad \text{a.s.}$

If $F \in \mathcal{F}_T$, then (check!) $F \cap \{T \leq k\} \in \mathcal{F}_{T \wedge k}$, so that, by $(*)$,

$(**)$ $\mathsf{E}(M_\infty; F \cap \{T \leq k\}) = \mathsf{E}(M_{T \wedge k}; F \cap \{T \leq k\}) = \mathsf{E}(M_T; F \cap \{T \leq k\}).$

We can (and do) restrict attention to the case when $M_\infty \geq 0$, whence $M_n = \mathsf{E}(M_\infty | \mathcal{F}_n) \geq 0$ for all n. Then, on letting $k \uparrow \infty$ in $(**)$ and using (MON), we obtain
$$\mathsf{E}(M_\infty; F \cap \{T < \infty\}) = \mathsf{E}(M_T; F \cap \{T < \infty\}).$$

However, the fact that
$$\mathsf{E}(M_\infty; F \cap \{T = \infty\}) = \mathsf{E}(M_T; F \cap \{T = \infty\}).$$

is tautological. Hence $\mathsf{E}(M_\infty; F) = \mathsf{E}(M_T; F)$.   □

Corollary 2 now follows from the Tower Property, and Corollary 1 follows from Corollary 2!

A14.4. The result for UI submartingales

A UI submartingale X has Doob decomposition
$$X = X_0 + M + A,$$

where (**Exercise:** explain why!) $\mathsf{E}(A_\infty) < \infty$ and M is UI. Hence, if T is a stopping time, then, almost surely,

$$\begin{aligned}
\mathsf{E}(X_\infty | \mathcal{F}_T) &= X_0 + \mathsf{E}(M_\infty | \mathcal{F}_T) + \mathsf{E}(A_\infty | \mathcal{F}_T) \\
&= X_0 + M_T + \mathsf{E}(A_\infty | \mathcal{F}_T) \\
&\geq X_0 + M_T + \mathsf{E}(A_T | \mathcal{F}_T) \\
&= X_T.
\end{aligned}$$

Chapter A16
Appendix to Chapter 16

A16.1. Differentiation under the integral sign

Before stating our theorem on this topic, let us examine the type of application we need in Section 16.3. Suppose that X is a RV such that $\mathsf{E}(|X|) < \infty$ and that $h(t, x) = ixe^{itx}$. (We can treat the real and imaginary parts of h separately.) Note that if $[a, b]$ is a subinterval of $\mathbf{R}$, then the variables $\{h(t, X) : t \in [a, b]\}$ are *dominated* by $|X|$, and so are UI. In the theorem, we shall have

$$\mathsf{E}H(t, X) = \varphi_X(t) - \varphi_X(a), \qquad t \in [a, b],$$

and we can conclude that $\varphi'_X(t)$ exists and equals $\mathsf{E}h(t, X)$.

THEOREM

Let X be a RV carried by $(\Omega, \mathcal{F}, \mathsf{P})$.

Suppose that $a, b \in \mathbf{R}$ with $a < b$, and that

$$h : [a, b] \times \mathbf{R} \to \mathbf{R}$$

has the properties:

(i) $t \mapsto h(t, x)$ *is continuous in t for every x in $\mathbf{R}$,*

(ii) $x \mapsto h(t, x)$ *is $\mathcal{B}$-measurable for every t in $[a, b]$,*

(iii) *the variables $\{h(t, X) : t \in [a, b]\}$ are UI.*

Then

(a) $t \mapsto \mathsf{E}h(t, X)$ *is continuous on $[a, b]$,*

(b) h *is $\mathcal{B}[a, b] \times \mathcal{B}$-measurable,*

(c)   *if $H(t,x) := \int_a^t h(s,x)ds$ for $a \le t \le b$, then for $t \in (a,b)$,*

$$\frac{d}{dt}\mathsf{E}H(t,X) \quad \text{exists and equals} \quad \mathsf{E}h(t,X).$$

Proof of (a). Since we need only consider the 'sequential case $t_n \to t$', result (a) follows immediately from Theorem 13.7.   □

Proof of (b). Define $\delta_n := 2^{-n}(b-a)$, $D_n := (a + \delta \mathbf{Z}^+) \cap [a,b]$,

$$\tau_n(t) := \inf\{\tau \in D_n : \tau \ge t\}, \qquad t \in [a,b],$$
$$h_n(t,x) := h(\tau_n(t),x), \qquad t \in [a,b], \; x \in \mathbf{R}.$$

Then, for $B \in \mathcal{B}$,

$$h_n^{-1}(B) = \bigcup_{\tau \in D_n} (([\tau, \tau + \delta) \cap [a,b]) \times \{x : h(\tau,x) \in B\}),$$

so that h_n is $\mathcal{B}[a,b] \times \mathcal{B}$-measurable. Since $h_n \to h$ on $[a,b] \times \mathbf{R}$, result (b) follows.   □

Proof of (c). For $\Gamma \subseteq [a,b] \times \mathbf{R}$, define

$$\alpha(\Gamma) := \{(t,\omega) \in [a,b] \times \Omega : (t, X(\omega)) \in \Gamma\}.$$

If $\Gamma = A \times C$ where $A \in \mathcal{B}[a,b]$ and $C \in \mathcal{B}$, then

$$\alpha(\Gamma) = A \times (X^{-1}C) \in \mathcal{B}[a,b] \times \mathcal{F}.$$

It is now clear that the class of Γ for which $\alpha(\Gamma)$ is an element of $\mathcal{B}[a,b] \times \mathcal{F}$ is a σ-algebra containing $\mathcal{B}[a,b] \times \mathcal{B}$. The point of all this is that

(∗) $(t,\omega) \mapsto h(t, X(\omega))$ is $\mathcal{B} \times \mathcal{F}$- measurable

since for $B \in \mathcal{B}$, $\{(t,\omega) : h(t,X(\omega)) \in B\}$ is $\alpha(h^{-1}B)$. (Yes, I know, we could have obtained (∗) more directly using the h_n's, but it is good to have other methods.) Since the family $\{h(t,X) : t \ge 0\}$ is UI, it is bounded in $\mathcal{L}^1$, whence

$$\int_a^b \mathsf{E}|h(t,X)|dt < \infty.$$

Fubini's Theorem now implies that, for $a \le t \le b$,

$$\int_a^t \mathsf{E}h(s,X)ds = \mathsf{E}\int_a^t h(s,X)ds = \mathsf{E}H(t,X),$$

and part (c) now follows.   □

Chapter E
Exercises

Starred exercises are more tricky. The first number in an exercise gives a rough indication of which **chapter** it depends on. 'G' stands for 'a bit of gumption is all that's necessary'. A number of exercises may also be found in the main text. Some are repeated here. We begin with an

Antidote to measure-theoretic material – *just for fun, though the point that probability is more than mere measure theory needs hammering home.*

EG.1. Two points are chosen at random on a line AB, each point being chosen according to the uniform distribution on AB, and the choices being made independently of each other. The line AB may now be regarded as divided into three parts. What is the probability that they may be made into a triangle?

EG.2. Planet X is a ball with centre O. Three spaceships A, B and C land at random on its surface, their positions being independent and each uniformly distributed on the surface. Spaceships A and B can communicate directly by radio if $\angle AOB < 90°$. Show that the probability that they can keep in touch (with, for example, A communicating with B via C if necessary) is $(\pi + 2)/(4\pi)$.

EG.3. Let G be the free group with two generators a and b. Start at time 0 with the unit element 1, the empty word. At each second multiply the current word on the right by one of the four elements a, a^{-1}, b, b^{-1}, choosing each with probability 1/4 (independently of previous choices). The choices

$$a, a, b, a^{-1}, a, b^{-1}, a^{-1}, a, b$$

at times 1 to 9 will produce the reduced word aab of length 3 at time 9. Prove that the probability that the reduced word 1 ever occurs at a positive time is 1/3, and explain why it is intuitively clear that (almost surely)

$$(\text{length of reduced word at time } n)/n \to \tfrac{1}{2}.$$

EG.4.* (Continuation) Suppose now that the elements a, a^{-1}, b, b^{-1} are chosen instead with respective probabilities $\alpha, \alpha, \beta, \beta$, where $\alpha > 0, \beta > 0, \alpha + \beta = \frac{1}{2}$. Prove that the conditional probability that the reduced word 1 ever occurs at a positive time, *given that* the element a is chosen at time 1, is the unique root $x = r(\alpha)$ (say) in $(0,1)$ of the equation

$$3x^3 + (3 - 4\alpha^{-1})x^2 + x + 1 = 0.$$

As time goes on, (it is almost surely true that) more and more of the reduced word becomes fixed, so that a final word is built up. If in the final word, the symbols a and a^{-1} are both replaced by A and the symbols b and b^{-1} are both replaced by B, show that the sequence of A's and B's obtained is a Markov chain on $\{A, B\}$ with (for example)

$$p_{AA} = \frac{\alpha(1 - x)}{\alpha(1 - x) + 2\beta(1 - y)},$$

where $y = r(\beta)$. What is the (almost sure) limiting proportion of occurrence of the symbol a in the final word? (*Note.* This result was used by Professor Lyons of Edinburgh to solve a long-standing problem in potential theory on Riemannian manifolds.)

Algebras, etc.

E1.1. 'Probability' for subsets of **N**

Let $V \subseteq \mathbf{N}$. Say that V has (Cesàro) density $\gamma(V)$ and write $V \in \text{CES}$ if

$$\gamma(V) := \lim \frac{\#(V \cap \{1, 2, 3, \ldots, n\})}{n}$$

exists. Give an example of sets V_1 and V_2 in CES for which $V_1 \cap V_2 \notin \text{CES}$. Thus, CES is not an algebra.

Independence

E4.1. Let $(\Omega, \mathcal{F}, \mathbf{P})$ be a probability triple. Let $\mathcal{I}_1, \mathcal{I}_2$ and $\mathcal{I}_3$ be three π-systems on Ω such that, for $k = 1, 2, 3$,

$$\mathcal{I}_k \subseteq \mathcal{F} \text{ and } \Omega \in \mathcal{I}_k.$$

Prove that if
$$\mathbf{P}(I_1 \cap I_2 \cap I_3) = \mathbf{P}(I_1)\mathbf{P}(I_2)\mathbf{P}(I_3)$$
whenever $I_k \in \mathcal{I}_k$ ($k = 1, 2, 3$), then $\sigma(\mathcal{I}_1), \sigma(\mathcal{I}_2), \sigma(\mathcal{I}_3)$ are independent. Why did we require that $\Omega \in \mathcal{I}_k$?

E4.2. Let $s > 1$, and define $\zeta(s) := \sum_{n \in \mathbf{N}} n^{-s}$, as usual. Let X and Y be independent N-valued random variables with

$$\mathbf{P}(X = n) = \mathbf{P}(Y = n) = n^{-s}/\zeta(s).$$

Prove that the events $(E_p : p \text{ prime})$, where $E_p = \{X \text{ is divisible by } p\}$, are independent. Explain Euler's formula

$$1/\zeta(s) = \prod_p (1 - 1/p^s)$$

probabilistically. Prove that

$$\mathbf{P}(\text{no square other than 1 divides } X) = 1/\zeta(2s).$$

Let H be the highest common factor of X and Y. Prove that

$$\mathbf{P}(H = n) = n^{-2s}/\zeta(2s).$$

E4.3. Let $X_1, X_2, \ldots$ be independent random variables with the same *continuous* distribution function. Let $E_1 := \Omega$, and, for $n \geq 2$, let

$$E_n := \{X_n > X_m, \forall m < n\} = \{\text{a 'Record' occurs at time } n\}.$$

Convince yourself and your tutor that the events $E_1, E_2, \ldots$ are independent, with $\mathbf{P}(E_n) = 1/n$. (Please see Note at the end of this Chapter.)

Borel-Cantelli Lemmas

E4.4. Suppose that a coin with probability p of heads is tossed repeatedly. Let A_k be the event that a sequence of k (or more) consecutive heads occurs amongst tosses numbered $2^k, 2^k + 1, 2^k + 2, \ldots, 2^{k+1} - 1$. Prove that

$$\mathbf{P}(A_k, \text{ i.o.}) = \begin{cases} 1 & \text{if } p \geq \frac{1}{2}, \\ 0 & \text{if } p < \frac{1}{2}. \end{cases}$$

Hint. Let E_i be the event that there are k consecutive heads beginning at toss numbered $2^k + (i-1)k$. Now make a simple use of the inclusion-exclusion formulae (Lemma 1.9).

E4.5. Prove that if G is a random variable with the normal N(0,1) distribution, then, for $x > 0$,

$$\mathbb{P}(G > x) = \frac{1}{\sqrt{2\pi}} \int_x^\infty e^{-\frac{1}{2}y^2}\, dy \leq \frac{1}{x\sqrt{2\pi}} e^{-\frac{1}{2}x^2}.$$

Let $X_1, X_2, \ldots$ be a sequence of independent N(0,1) variables. Prove that, with probability 1, $L \leq 1$, where

$$L := \limsup(X_n/\sqrt{2\log n}).$$

(*Harder.* Prove that $\mathbb{P}(L = 1) = 1$.) [*Hint.* See Section 14.8.]

Let $S_n := X_1 + X_2 + \cdots + X_n$. Recall that $S_n/\sqrt{n}$ has the N(0,1) distribution. Prove that

$$\mathbb{P}(|S_n| < 2\sqrt{n\log n},\ \text{ev}) = 1.$$

Note that this implies the Strong Law: $\mathbb{P}(S_n/n \to 0) = 1$.

Remark. The *Law of the Iterated Logarithm* states that

$$\mathbb{P}\left(\limsup \frac{S_n}{\sqrt{2n\log\log n}} = 1\right) = 1.$$

Do not attempt to prove this now! See Section 14.7.

E4.6. Converse to SLLN

Let Z be a non-negative RV. Let Y be the integer part of Z. Show that

$$Y = \sum_{n\in\mathbb{N}} I_{\{Z\geq n\}},$$

and deduce that

$$(*) \qquad \sum_{n\in\mathbb{N}} \mathbb{P}[Z \geq n] \leq \mathbb{E}(Z) \leq 1 + \sum_{n\in\mathbb{N}} \mathbb{P}[Z \geq n].$$

Let (X_n) be a sequence of IID RVs (independent, identically distributed random variables) with $\mathbb{E}(|X_n|) = \infty, \forall n$. Prove that

$$\sum_n \mathbb{P}[|X_n| > kn] = \infty \ (k \in \mathbb{N}) \quad \text{and} \quad \limsup \frac{|X_n|}{n} = \infty, \quad \text{a.s.}$$

Deduce that if $S_n = X_1 + X_2 + \cdots + X_n$, then

$$\limsup \frac{|S_n|}{n} = \infty, \quad \text{a.s.}$$

E4.7. What's fair about a fair game?

Let $X_1, X_2, \ldots$ be independent RVs such that

$$X_n = \begin{cases} n^2 - 1 & \text{with probability } n^{-2} \\ -1 & \text{with probability } 1 - n^{-2}. \end{cases}$$

Prove that $\mathbf{E}(X_n) = 0$, $\forall n$, but that if $S_n = X_1 + X_2 + \cdots + X_n$, then

$$\frac{S_n}{n} \to -1, \quad \text{a.s.}$$

E4.8*. Blackwell's test of imagination

This exercise assumes that you are familiar with continuous-parameter Markov chains with two states.

For each $n \in \mathbb{N}$, let $X^{(n)} = \{X^{(n)}(t) : t \geq 0\}$ be a Markov chain with state-space the two-point set $\{0, 1\}$ with Q-matrix

$$Q^{(n)} = \begin{pmatrix} -a_n & a_n \\ b_n & -b_n \end{pmatrix}, \quad a_n, b_n > 0,$$

and transition function $P^{(n)}(t) = \exp(tQ^{(n)})$. Show that, for every t,

$$p_{00}^{(n)}(t) \geq b_n/(a_n + b_n), \quad p_{01}^{(n)}(t) \leq a_n/(a_n + b_n).$$

The processes $(X^{(n)} : n \in \mathbb{N})$ are independent and $X^{(n)}(0) = 0$ for every n. Each $X^{(n)}$ has right-continuous paths.

Suppose that $\sum a_n = \infty$ *and* $\sum a_n/b_n < \infty$.

Prove that if t is a fixed time then

(*) $\mathbf{P}\{X^{(n)}(t) = 1 \text{ for infinitely many } n\} = 0.$

Use Weierstrass's M-test to show that $\sum_n \log p_{00}^{(n)}(t)$ is *uniformly* convergent on $[0, 1]$, and deduce that

$$\mathbf{P}\{X^{(n)}(t) = 0 \text{ for ALL } n\} \to 1 \quad \text{as } t \downarrow 0.$$

Prove that

$$\mathbf{P}\{X^{(n)}(s) = 0, \forall s \leq t, \forall n\} = 0 \text{ for every } t > 0$$

and discuss with your tutor why it is almost surely true that

(∗∗)   within every non-empty time interval, infinitely many of the $X^{(n)}$ chains jump.

Now imagine the whole behaviour.

Notes. Almost surely, the process $X = (X^{(n)})$ spends *almost* all its time in the countable subset of $\{0,1\}^{\mathbf{N}}$ consisting of sequences with only finitely many 1's. This follows from (∗) and Fubini's Theorem 8.2. However, it is a.s. true that X visits uncountable points of $\{0,1\}^{\mathbf{N}}$ during every non-empty time interval. This follows from (∗∗) and the Baire category theorem A1.12. By using much deeper techniques, one can show that for certain choices of (a_n) and (b_n), X will almost certainly visit *every* point of $\{0,1\}^{\mathbf{N}}$ uncountably often within a finite time.

Tail σ-algebras

E4.9. Let $Y_0, Y_1, Y_2, \ldots$ be independent random variables with

$$\mathbf{P}(Y_n = +1) = \mathbf{P}(Y_n = -1) = \tfrac{1}{2}, \quad \forall n.$$

For $n \in \mathbf{N}$, define

$$X_n := Y_0 Y_1 \ldots Y_n.$$

Prove that the variables $X_1, X_2, \ldots$ are independent. Define

$$\mathcal{Y} := \sigma(Y_1, Y_2, \ldots), \quad \mathcal{T}_n := \sigma(X_r : r > n).$$

Prove that

$$\mathcal{L} := \bigcap_n \sigma(\mathcal{Y}, \mathcal{T}_n) \neq \sigma\left(\mathcal{Y}, \bigcap_n \mathcal{T}_n\right) =: \mathcal{R}.$$

Hint. Prove that $Y_0 \in m\mathcal{L}$ and that Y_0 is independent of $\mathcal{R}$.

E4.10. *Star Trek, 2*

See E10.11, which you can do now.

Dominated-Convergence Theorem

E5.1. Let $S := [0,1]$, $\Sigma := \mathcal{B}(S)$, $\mu := $ Leb. Define $f_n := nI_{(0,1/n)}$. Prove that $f_n(s) \to 0$ for every s in S, but that $\mu(f_n) = 1$ for every n. Draw a picture of $g := \sup_n |f_n|$, and show that $g \notin \mathcal{L}^1(S, \Sigma, \mu)$.

Inclusion-Exclusion Formulae

E5.2. Prove the inclusion-exclusion formulae and inequalities of Section 1.9 by considering integrals of indicator functions.

The Strong Law

E7.1. Inverting Laplace transforms

Let f be a bounded continuous function on $[0, \infty)$. The Laplace transform of f is the function L on $(0, \infty)$ defined by

$$L(\lambda) := \int_0^\infty e^{-\lambda x} f(x) dx.$$

Let $X_1, X_2, \ldots$ be independent RVs each with the exponential distribution of rate λ, so $P[X > x] = e^{-\lambda x}$, $E(X) = \frac{1}{\lambda}$, $\text{Var}(X) = \frac{1}{\lambda^2}$. Show that

$$(-1)^{n-1} \frac{\lambda^n L^{(n-1)}(\lambda)}{(n-1)!} = Ef(S_n),$$

where $S_n = X_1 + X_2 + \cdots + X_n$, and $L^{(n-1)}$ denotes the $(n-1)^{th}$ derivative of L. Prove that f may be recovered from L as follows: for $y > 0$,

$$f(y) = \lim_{n \uparrow \infty} (-1)^{n-1} \frac{(n/y)^n L^{(n-1)}(n/y)}{(n-1)!}.$$

E7.2. The uniform distribution on the sphere $S^{n-1} \subseteq \mathbf{R}^n$

As usual, write $S^{n-1} = \{x \in \mathbf{R}^n : |x| = 1\}$. You may assume that there is a unique probability measure ν^{n-1} on $(S^{n-1}, \mathcal{B}(S^{n-1}))$ such that $\nu^{n-1}(A) = \nu^{n-1}(HA)$ for every orthogonal $n \times n$ matrix H and every A in $\mathcal{B}(S^{n-1})$.

Prove that if $\mathbf{X}$ is a vector in $\mathbf{R}_n$, the components of which are independent $N(0,1)$ variables, then for every orthogonal $n \times n$ matrix H, the vector $H\mathbf{X}$ has the same property. Deduce that $\mathbf{X}/|\mathbf{X}|$ has law ν^{n-1}.

Let $Z_1, Z_2, \ldots$ be independent $N(0,1)$ variables and define

$$R_n = (Z_1^2 + Z_2^2 + \cdots + Z_n^2)^{\frac{1}{2}}.$$

Prove that $R_n/\sqrt{n} \to 1$, a.s.

Combine these ideas to prove a rather striking fact which relates the normal distribution to the 'infinite-dimensional' sphere and which is important both for Brownian motion and for Fock-space constructions in quantum mechanics:

If, for each n, $(Y_1^{(n)}, Y_2^{(n)}, \ldots, Y_n^{(n)})$ is a point chosen on S^{n-1} according to the distribution ν^{n-1}, then

$$\lim_{n \to \infty} \mathbf{P}(\sqrt{n}Y_1^{(n)} \le x) = \Phi(x) = \frac{1}{\sqrt{2\pi}} \int_{-\infty}^{x} e^{-y^2/2} dy,$$

$$\lim_{n \to \infty} \mathbf{P}(\sqrt{n}Y_1^{(n)} \le x_1; \sqrt{n}Y_2^{(n)} \le x_2) = \Phi(x_1)\Phi(x_2).$$

Hint. $\mathbf{P}(Y_1^{(n)} \le u) = \mathbf{P}(X_1/R_n \le u)$.

Conditional Expectation

E9.1. Prove that if $\mathcal{G}$ is a sub-σ-algebra of $\mathcal{F}$ and if $X \in \mathcal{L}^1(\Omega, \mathcal{F}, \mathbf{P})$ and if $Y \in \mathcal{L}^1(\Omega, \mathcal{G}, \mathbf{P})$ and

(*) $\mathbf{E}(X; G) = \mathbf{E}(Y; G)$

for every G in a π-system which contains Ω and generates $\mathcal{G}$, then (*) holds for every G in $\mathcal{G}$.

E9.2. Suppose that $X, Y \in \mathcal{L}^1(\Omega, \mathcal{F}, \mathbf{P})$ and that

$$\mathbf{E}(X|Y) = Y, \quad \text{a.s.}, \qquad \mathbf{E}(Y|X) = X, \quad \text{a.s.}$$

Prove that $\mathbf{P}(X = Y) = 1$.

Hint. Consider $\mathbf{E}(X - Y; X > c, Y \le c) + \mathbf{E}(X - Y; X \le c, Y \le c)$.

Martingales

E10.1. Pólya's urn

At time 0, an urn contains 1 black ball and 1 white ball. At each time $1, 2, 3, \ldots$, a ball is chosen at random from the urn and is replaced together with a new ball of the same colour. Just after time n, there are therefore $n + 2$ balls in the urn, of which $B_n + 1$ are black, where B_n is the number of black balls chosen by time n.

Let $M_n = (B_n + 1)/(n + 2)$, the proportion of black balls in the urn just after time n. Prove that (relative to a natural filtration which you should specify) M is a martingale.

Prove that $\mathbf{P}(B_n = k) = (n + 1)^{-1}$ for $0 \le k \le n$. What is the distribution of Θ, where $\Theta := \lim M_n$?

Prove that for $0 < \theta < 1$,

$$N_n^\theta := \frac{(n+1)!}{B_n!(n-B_n)!}\,\theta^{B_n}(1-\theta)^{n-B_n}$$

defines a martingale N^θ.

(*Continued at E10.8.*)

E10.2. Martingale formulation of Bellman's Optimality Principle

Your winnings per unit stake on game n are ε_n, where the ε_n are IID RVs with

$$\mathbf{P}(\varepsilon_n = +1) = p, \mathbf{P}(\varepsilon_n = -1) = q, \quad \text{where} \quad \tfrac{1}{2} < p = 1 - q < 1.$$

Your stake C_n on game n must lie between 0 and Z_{n-1}, where Z_{n-1} is your fortune at time $n-1$. Your object is to maximize the expected 'interest rate' $\mathbf{E}\log(Z_N/Z_0)$, where N is a given integer representing the length of the game, and Z_0, your fortune at time 0, is a given constant. Let $\mathcal{F}_n = \sigma(\varepsilon_1, \ldots, \varepsilon_n)$ be your 'history' up to time n. Show that if C is any (previsible) strategy, then $\log Z_n - n\alpha$ is a *super*martingale, where α denotes the '*entropy*'

$$\alpha = p\log p + q\log q + \log 2,$$

so that $\mathbf{E}\log(Z_N/Z_0) \le N\alpha$, but that, for a certain strategy, $\log Z_n - n\alpha$ is a *martingale*. What is the best strategy?

E10.3. Stopping times

Suppose that S and T are stopping times (relative to $(\Omega, \mathcal{F}, \{\mathcal{F}_n\})$). Prove that $S \wedge T \; (:= \min(S,T))$, $S \vee T (:= \max(S,T))$ and $S + T$ are stopping times.

E10.4. Let S and T be stopping times with $S \le T$. Define the process $1_{(S,T]}$ with parameter set $\mathbf{N}$ via

$$1_{(S,T]}(n,\omega) := \begin{cases} 1 & \text{if } S(\omega) < n \le T(\omega), \\ 0 & \text{otherwise.} \end{cases}$$

Prove that $1_{(S,T]}$ is previsible, and deduce that if X is a supermartingale, then

$$\mathbf{E}(X_{T \wedge n}) \le \mathbf{E}(X_{S \wedge n}), \quad \forall n.$$

E10.5. *'What always stands a reasonable chance of happening will (almost surely) happen – sooner rather than later.'*

Suppose that T is a stopping time such that for some $N \in \mathbf{N}$ and some $\varepsilon > 0$, we have, for every n:

$$\mathbf{P}(T \leq n + N | \mathcal{F}_n) > \varepsilon, \quad \text{a.s.}$$

Prove by induction using $\mathbf{P}(T > kN) = \mathbf{P}(T > kN; T > (k-1)N)$ that for $k = 1, 2, 3, \ldots$

$$\mathbf{P}(T > kN) \leq (1 - \varepsilon)^k.$$

Show that $\mathbf{E}(T) < \infty$.

E10.6. ABRACADABRA

At each of times $1,2,3,\ldots$, a monkey types a capital letter at random, the sequence of letters typed forming an IID sequence of RVs each chosen uniformly from amongst the 26 possible capital letters.

Just before **each** time $n = 1, 2, \ldots$, a new gambler arrives on the scene. He bets $ 1 that

the nth letter will be A.

If he loses, he leaves. If he wins, he receives $ 26 all of which he bets on the event that

the $(n+1)^{\text{th}}$ letter will be B.

If he loses, he leaves. If he wins, he bets his whole current fortune of $ 26^2 that

the $(n+2)^{\text{th}}$ letter will be R

and so on through the ABRACADABRA sequence. Let T be the first time by which the monkey has produced the consecutive sequence ABRA-CADABRA. Explain why martingale theory makes it intuitively obvious that

$$\mathbf{E}(T) = 26^{11} + 26^4 + 26$$

and use result 10.10(c) to prove this. (See Ross (1983) for other such applications.)

E10.7. Gambler's Ruin

Suppose that $X_1, X_2, \ldots$ are IID RVs with

$$\mathbf{P}[X = +1] = p, \quad \mathbf{P}[X = -1] = q, \quad \text{where} \quad 0 < p = 1 - q < 1,$$

and $p \neq q$. Suppose that a and b are integers with $0 < a < b$. Define

$$S_n := a + X_1 + \cdots + X_n, \quad T := \inf\{n : S_n = 0 \quad \text{or} \quad S_n = b\}.$$

Let $\mathcal{F}_n = \sigma(X_1, \ldots, X_n)$ ($\mathcal{F}_0 = \{\emptyset, \Omega\}$). Explain why T satisfies the conditions in Question E10.5. Prove that

$$M_n := (\frac{q}{p})^{S_n} \quad \text{and} \quad N_n = S_n - n(p-q)$$

define martingales M and N. Deduce the values of $\mathbb{P}(S_T = 0)$ and $\mathbb{E}(T)$.

E10.8. Bayes' urn

A random number Θ is chosen uniformly between 0 and 1, and a coin with probability Θ of heads is minted. The coin is tossed repeatedly. Let B_n be the number of heads in n tosses. Prove that (B_n) has exactly the same probabilistic structure as the (B_n) sequence in (E10.1) on Polya's urn. Prove that N_n^θ is a regular conditional pdf of Θ given $B_1, B_2, \ldots, B_n$.

(Continued at E18.5.)

E10.9. Show that if X is a non-negative supermartingale and T is a stopping time, then

$$\mathbb{E}(X_T; T < \infty) \leq \mathbb{E}(X_0).$$

(*Hint.* Recall Fatou's Lemma.) Deduce that $c\,\mathbb{P}(\sup_n X_n \geq c) \leq \mathbb{E}(X_0)$.

E10.10*. The 'Star-ship *Enterprise*' Problem

The control system on the star-ship *Enterprise* has gone wonky. All that one can do is to set a distance to be travelled. The spaceship will then move that distance in a randomly chosen direction, then stop. The object is to get into the Solar System, a ball of radius r. Initially, the *Enterprise* is at a distance $R_0(> r)$ from the Sun.

Let R_n be the distance from Sun to *Enterprise* after n 'space-hops'. Use Gauss's theorems on potentials due to spherically-symmetric charge distributions to show that whatever strategy is adopted, $1/R_n$ is a *supermartingale*, and that for any strategy which always sets a distance no greater than that from Sun to *Enterprise*, $1/R_n$ is a *martingale*. Use (E10.9) to show that

$$\mathbb{P}[\textit{Enterprise gets into Solar System}] \leq r/R_0.$$

For each $\varepsilon > 0$, you can choose a strategy which makes this probability greater than $(r/R_0) - \varepsilon$. What kind of strategy will this be?

E10.11*. *Star Trek*, 2. 'Captain's Log ...

Mr Spock and Chief Engineer Scott have modified the control system so that the *Enterprise* is confined to move for ever in a fixed plane passing through the Sun. However, the next 'hop-length' is now automatically set to be the current distance to the Sun ('next' and 'current' being updated in the obvious way). Spock is muttering something about logarithms and random walks, but I wonder whether it is (almost) certain that we will get into the Solar System sometime ... '

Hint. Let $X_n := \log R_n - \log R_{n-1}$. Prove that $X_1, X_2, \ldots$ is an IID sequence of variables each of mean 0 and finite variance σ^2 (say), where $\sigma > 0$. let

$$S_n := X_1 + X_2 + \cdots + X_n.$$

Prove that if α is a fixed positive number, then

$$\mathbf{P}[\inf_n S_n = -\infty] \geq \mathbf{P}[S_n \leq -\alpha\sigma\sqrt{n}, \text{i.o.}]$$
$$\geq \limsup \mathbf{P}[S_n \leq -\alpha\sigma\sqrt{n}] = \Phi(-\alpha) > 0.$$

(Use the Central Limit Theorem.) Prove that the event $\{\inf_n S_n = -\infty\}$ is in the tail σ-algebra of the (X_n) sequence.

E12.1. Branching Process

A branching process $Z = \{Z_n : n \geq 0\}$ is constructed in the usual way. Thus, a family $\left\{X_k^{(n)} : n, k \geq 1\right\}$ of IID $\mathbf{Z}^+$-valued random variables is supposed given. We define $Z_0 := 1$ and then define recursively:

$$Z_{n+1} := X_1^{(n+1)} + \cdots + X_{Z_n}^{(n+1)} \qquad (n \geq 0).$$

Assume that if X denotes any one of the $X_k^{(n)}$, then

$$\mu := \mathbf{E}(X) < \infty \quad \text{and} \quad 0 < \sigma^2 := \text{Var}(X) < \infty.$$

Prove that $M_n := Z_n/\mu^n$ defines a martingale M relative to the filtration $\mathcal{F}_n = \sigma(Z_0, Z_1, \ldots, Z_n)$. Show that

$$\mathbf{E}\left(Z_{n+1}^2 | \mathcal{F}_n\right) = \mu^2 Z_n^2 + \sigma^2 Z_n$$

and deduce that M is bounded in $\mathcal{L}^2$ if and only if $\mu > 1$. Show that when $\mu > 1$,

$$\text{Var}(M_\infty) = \sigma^2 \{\mu(\mu - 1)\}^{-1}.$$

E12.2. Use of Kronecker's Lemma

Let $E_1, E_2, \ldots$ be independent events with $\mathbb{P}(E_n) = 1/n$. Let $Y_i = I_{E_i}$. Prove that $\sum \left(Y_k - \frac{1}{k}\right) / \log k$ converges a.s., and use Kronecker's Lemma to deduce that

$$\frac{N_n}{\log n} \longrightarrow 1, \quad \text{a.s.,}$$

where $N_n := Y_1 + \cdots + Y_n$. An interesting application is to E4.3, when N_n becomes the *number of records* by time n.

E12.3. *Star Trek, 3*

Prove that if the strategy in E10.11 is (in the obvious sense) employed – and for ever – *in $\mathbb{R}^3$ rather than in $\mathbb{R}^2$*, then

$$\sum R_n^{-2} < \infty, \quad \text{a.s.,}$$

where R_n is the distance from the *Enterprise* to the Sun at time n.

Note. It should be obvious which result plays the key rôle here, but you should try to make your argument fully rigorous.

Uniform Integrability

E13.1. Prove that a class $\mathcal{C}$ of RVs is UI if and only if both of the following conditions (i) and (ii) hold:
  (i) $\mathcal{C}$ is bounded in $\mathcal{L}^1$, so that $A := \sup\{\mathbb{E}(|X|) : X \in \mathcal{C}\} < \infty$,
  (ii) for every $\varepsilon > 0$, $\exists \delta > 0$ such that if $F \in \mathcal{F}$, $\mathbb{P}(F) < \delta$ and $X \in \mathcal{C}$, then $\mathbb{E}(|X|; F) < \varepsilon$.
Hint for 'if'. For $X \in \mathcal{C}$, $\mathbb{P}(|X| > K) \leq K^{-1}A$.
Hint for 'only if'. $\mathbb{E}(|X|; F) \leq \mathbb{E}(|X|; |X| > K) + K\mathbb{P}(F)$.

E13.2. Prove that if $\mathcal{C}$ and $\mathcal{D}$ are UI classes of RVs, and if we define

$$\mathcal{C} + \mathcal{D} := \{X + Y : X \in \mathcal{C}, Y \in \mathcal{D}\},$$

then $\mathcal{C} + \mathcal{D}$ is UI. *Hint.* One way to prove this is to use E13.1.

E13.3. Let $\mathcal{C}$ be a UI family of RVs. Say that $Y \in \mathcal{D}$ if for some $X \in \mathcal{C}$ and some sub-σ-algebra $\mathcal{G}$ of $\mathcal{F}$, we have $Y = \mathbb{E}(X|\mathcal{G})$, a.s. Prove that $\mathcal{D}$ is UI.

E14.1. Hunt's Lemma

Suppose that (X_n) is a sequence of RVs such that $X := \lim X_n$ exists a.s. and that (X_n) is dominated by Y in $\left(\mathcal{L}^1\right)^+$:

$$|X_n(\omega)| \leq Y(\omega), \quad \forall(n, \omega), \quad \text{and} \quad \mathbb{E}(Y) < \infty.$$

Let $\{\mathcal{F}_n\}$ be any filtration. Prove that

$$\mathbb{E}(X_n|\mathcal{F}_n) \to \mathbb{E}(X|\mathcal{F}_\infty) \quad \text{a.s.}$$

Hint. Let $Z_m := \sup_{r \ge m} |X_r - X|$. Prove that $Z_m \to 0$ a.s. and in $\mathcal{L}^1$. Prove that for $n \ge m$, we have, almost surely,

$$|\mathbb{E}(X_n|\mathcal{F}_n) - \mathbb{E}(X|\mathcal{F}_\infty)| \le |\mathbb{E}(X|\mathcal{F}_n) - \mathbb{E}(X|\mathcal{F}_\infty)| + \mathbb{E}(Z_m|\mathcal{F}_n).$$

E14.2. Azuma-Hoeffding Inequality

(a) Show that if Y is a RV with values in $[-c, c]$ and with $\mathbb{E}(Y) = 0$, then, for $\theta \in \mathbb{R}$,

$$\mathbb{E}e^{\theta Y} \le \cosh \theta c \le \exp\left(\frac{1}{2}\theta^2 c^2\right).$$

(b) Prove that if M is a martingale null at 0 such that for some sequence $(c_n : n \in \mathbb{N})$ of positive constants,

$$|M_n - M_{n-1}| \le c_n, \quad \forall n,$$

then, for $x > 0$,

$$\mathbb{P}\left(\sup_{k \le n} M_k \ge x\right) \le \exp\left(-\frac{1}{2}x^2 \bigg/ \sum_{k=1}^{n} c_k^2\right).$$

Hint for (a). Let $f(z) := \exp(\theta z)$, $z \in [-c, c]$. Then, since f is convex,

$$f(y) \le \frac{c - y}{2c}f(-c) + \frac{c + y}{2c}f(c).$$

Hint for (b). See the proof of (14.7,a).

Characteristic Functions

E16.1. Prove that

$$\lim_{T \uparrow \infty} \int_0^T x^{-1} \sin x \, dx = \pi/2$$

by integrating $\int z^{-1}e^{iz}dz$ around the contour formed by the 'upper' semi-circles of radii ε and T and the intervals $[-T, -\varepsilon]$ and $[\varepsilon, T]$.

E16.2. Prove that if Z has the U$[-1, 1]$ distribution, then

$$\varphi_Z(\theta) = (\sin \theta)/\theta,$$

and prove that there do not exist IID RVs X and Y such that

$$X - Y \sim U[-1, 1].$$

E16.3. Suppose that X has the Cauchy distribution, and let $\theta > 0$. By integrating $e^{i\theta z}/(1 + z^2)$ around the semicircle formed by $[-R, R]$ together with the 'upper' semicircle centre 0 of radius R, prove that $\varphi_X(\theta) = e^{-\theta}$. Show that $\varphi_X(\theta) = e^{-|\theta|}$ for all θ. Prove that if $X_1, X_2, \ldots X_n$ are IID RVs each with the standard Cauchy distribution, then $(X_1 + \cdots + X_n)/n$ also has the standard Cauchy distribution.

E16.4. Suppose that X has the standard normal N(0,1) distribution. Let $\theta > 0$. Consider $\int (2\pi)^{-\frac{1}{2}} \exp(-\frac{1}{2}z^2) dz$ around the rectangular contour

$$(-R - i\theta) \to (R - i\theta) \to R \to (-R) \to (-R - i\theta),$$

and prove that $\varphi_X(\theta) = \exp(-\frac{1}{2}\theta^2)$.

E16.5. Prove that if φ is the characteristic function of a RV X, then φ is *non-negative definite* in that for complex $c_1, c_2, \ldots, c_n$ and real $\theta_1, \theta_2, \ldots, \theta_n$,

$$\sum_j \sum_k c_j \bar{c}_k \varphi(\theta_j - \theta_k) \geq 0.$$

(*Hint.* Express LHS as the expectation of ... .) **Bochner's Theorem** says that φ is a characteristic function *if and only if* $\varphi(0) = 1$, φ is continuous, and φ is non-negative definite! (It is of course understood that here $\varphi : \mathbb{R} \to \mathbb{C}$.) E18.6 gives a simpler result in the same spirit.

E16.6. (a) Let $(\Omega, \mathcal{F}, \mathbf{P}) = ([0, 1], \mathcal{B}[0, 1], \text{Leb})$. What is the distribution of the RV Z, where $Z(\omega) := 2\omega - 1$? Let $\omega = \sum 2^{-n} R_n(\omega)$ be the binary expansion of ω. Let

$$U(\omega) = \sum_{\text{odd } n} 2^{-n} Q_n(\omega), \quad \text{where} \quad Q_n(\omega) = 2R_n(\omega) - 1.$$

Find a random variable V independent of U such that U and V are identically distributed and $U + \frac{1}{2}V$ is uniformly distributed on $[-1, 1]$.

(b) Now suppose that (on some probability triple) X and Y are IID RVs such that

$$X + \tfrac{1}{2}Y \quad \text{is uniformly distributed on} \quad [-1, 1].$$

Let φ be the CF of X. Calculate $\varphi(\theta)/\varphi(\frac{1}{4}\theta)$. Show that the distribution of X must be the same as that of U in part (a), and deduce that there exists a set $F \in \mathcal{B}[-1,1]$ such that $\mathrm{Leb}(F) = 0$ and $\mathbb{P}(X \in F) = 1$.

E18.1. (a) Suppose that $\lambda > 0$ and that (for $n > \lambda$)F_n is the DF associated with the Binomial distribution $B(n, \lambda/n)$. Prove (using CFs) that F_n converges weakly to F where F is the DF of the Poisson distribution with parameter λ.

(b) Suppose that $X_1, X_2, \ldots$ are IID RVs each with the density function $(1 - \cos x)/\pi x^2$ on $\mathbb{R}$. Prove that for $x \in \mathbb{R}$,

$$\lim_{n \to \infty} \mathbb{P}\left(\frac{X_1 + X_2 + \cdots + X_n}{n} \leq x\right) = \tfrac{1}{2} + \pi^{-1} \arc\tan x,$$

where $\arc\tan \in (-\frac{\pi}{2}, \frac{\pi}{2})$.

E18.2. Prove the Weak Law of Large Numbers in the following form. Suppose that $X_1, X_2, \ldots$ are IID RVs, each with the same distribution as X. Suppose that $X \in \mathcal{L}^1$ and that $\mathbb{E}(X) = \mu$. Prove by the use of CFs that the distribution of

$$A_n := n^{-1}(X_1 + \cdots + X_n)$$

converges weakly to the unit mass at μ. Deduce that

$$A_n \to \mu \quad \text{in probability.}$$

Of course, SLLN implies this Weak Law.

Weak Convergence for Prob$[0, 1]$

E18.3. Let X and Y be RVs taking values in $[0,1]$. Suppose that

$$\mathbb{E}(X^k) = \mathbb{E}(Y^k), \qquad k = 0, 1, 2, \ldots .$$

Prove that

   (i) $\mathbb{E}p(X) = \mathbb{E}p(Y)$ for every polynomial p,

   (ii) $\mathbb{E}f(X) = \mathbb{E}f(Y)$ for every continuous function f on $[0,1]$,

   (iii) $\mathbb{P}(X \leq x) = \mathbb{P}(Y \leq x)$ for every x in $[0,1]$.

Hint for (ii). Use the Weierstrass Theorem 7.4.

E18.4. Suppose that (F_n) is a sequence of DFs with

$$F_n(x) = 0 \text{ for } x < 0, \quad F_n(1) = 1, \qquad \text{for every } n.$$

Suppose that

(*) $m_k := \lim_n \int_{[0,1]} x^k dF_n$ exists for $k = 0, 1, 2, \ldots$.

Use the Helly-Bray Lemma and E18.3 to show that $F_n \overset{w}{\to} F$, where F is characterized by $\int_{[0,1]} x^k dF = m_k$, $\forall k$.

E18.5. Improving on E18.3: A Moment Inversion Formula

Let F be a distribution with $F(0-) = 0$ and $F(1) = 1$. Let μ be the associated law, and define

$$m_k := \int_{[0,1]} x^k dF(x).$$

Define

$$\Omega = [0,1] \times [0,1]^{\mathbf{N}}, \quad \mathcal{F} = \mathcal{B} \times \mathcal{B}^{\mathbf{N}}, \quad \mathbf{P} = \mu \times \mathrm{Leb}^{\mathbf{N}},$$
$$\Theta(\omega) = \omega_0, \quad H_k(\omega) = I_{[0,\omega_0]}(\omega_k).$$

This models the situation in which Θ is chosen with law μ, a coin with probability Θ of heads is then minted, and tossed at times $1, 2, \ldots$. See E10.8. The RV H_k is 1 if the kth toss produces heads, 0 otherwise. Define

$$S_n := H_1 + H_2 + \cdots + H_n.$$

By the Strong Law and Fubini's Theorem,

$$S_n/n \to \Theta, \quad \text{a.s.}$$

Define a map D on the space of real sequences $(a_n : n \in \mathbf{Z}^+)$ by setting

$$Da = (a_n - a_{n+1} : n \in \mathbf{Z}^+).$$

Prove that

(*) $F_n(x) := \sum_{i \leq nx} \binom{n}{i}(D^{n-i}m)_i \to F(x)$

at every point x of continuity of F.

E18.6* Moment Problem

Prove that if $(m_k : k \in \mathbf{Z}^+)$ is a sequence of numbers in $[0,1]$, then there exists a RV X with values in $[0,1]$ such that $\mathbf{E}(X^k) = m_k$ if and only if $m_0 = 1$ and

$$(D^r m)_s \geq 0 \qquad (r, s, \in \mathbf{Z}^+).$$

Hint. Define F_n via E18.5(*), and then verify that E18.4(*) holds. You can show that the moments $m_{k,n}$ of F_n satisfy

$$m_{n,0} = 1, \quad m_{n,1} = m_1, \quad m_{n,2} = m_2 + n^{-1}(m_1 - m_2), \quad \text{etc.}$$

You discover the algebra!

Weak Convergence for Prob[0, ∞)

E18.7. Using Laplace transforms instead of CFs

Suppose that F and G are DFs on **R** such that $F(0-) = G(0-) = 0$, and

$$\int_{[0,\infty)} e^{-\lambda x} dF(x) = \int_{[0,\infty)} e^{-\lambda x} dG(x), \qquad \forall \lambda \geq 0.$$

Note that the integral on LHS has a contribution $F(0)$ from $\{0\}$. Prove that $F = G$. [*Hint.* One could derive this from the idea in E7.1. However, it is easier to use E18.3, because we know that if X has DF F and Y has DF G, then

$$\mathsf{E}[(e^{-X})^n] = \mathsf{E}[(e^{-Y})^n], \qquad n = 0, 1, 2, \dots.]$$

Suppose that (F_n) is a sequence of distribution functions on **R** each with $F_n(0-) = 0$ and such that

$$L(\lambda) := \lim_n \int e^{-\lambda x} dF_n(x)$$

exists for $\lambda \geq 0$ and that L is continuous at 0. Prove that F_n is tight and that

$$F_n \xrightarrow{w} F \text{ where } \int e^{-\lambda x} dF(x) = L(\lambda), \qquad \forall \lambda \geq 0.$$

Modes of convergence

EA13.1. (a) Prove that $(X_n \to X, \text{ a.s.}) \Rightarrow (X_n \to X \text{ in prob})$.

Hint. See Section 13.5.

(b) Prove that $(X_n \to X \text{ in prob}) \nRightarrow (X_n \to X, \text{ a.s.})$.

Hint. Let $X_n = I_{E_n}$, where $E_1, E_2, \dots$ are independent events.

(c) Prove that if $\sum \mathsf{P}(|X_n - X| > \varepsilon) < \infty, \forall \varepsilon > 0$, then $X_n \to X$, a.s.

Hint. Show that the set $\{\omega : X_n(\omega) \nrightarrow X(\omega)\}$ may be written

$$\bigcup_{k \in \mathbf{N}} \{\omega : |X_n(\omega) - X(\omega)| > k^{-1} \text{ for infinitely many } n\}.$$

(d) Suppose that $X_n \to X$ in probability. Prove that there is a subsequence (X_{n_k}) of (X_n) such that $X_{n_k} \to X$, a.s.

Hint. Combine (c) with the 'diagonal principle'.

(e) Deduce from (a) and (d) that $X_n \to X$ in probability if and only if every subsequence of (X_n) contains a further subsequence which converges a.s. to X.

EA13.2. Recall that if ξ is a random variable with the standard normal N(0,1) distribution, then

$$\mathsf{E}e^{\lambda\xi} = \exp(\tfrac{1}{2}\lambda^2).$$

Suppose that $\xi_1, \xi_2, \ldots$ are IID RVs each with the N(0,1) distribution. Let $S_n = \sum_{k=1}^n \xi_k$, let $a, b \in \mathbf{R}$, and define

$$X_n = \exp(aS_n - bn).$$

Prove that

$$(X_n \to 0, \text{ a.s.}) \Leftrightarrow (b > 0)$$

but that for $r \geq 1$,

$$(X_n \to 0 \text{ in } \mathcal{L}^r) \Leftrightarrow (r < 2b/a^2).$$

Note on E4.3. Some students with a keen eye for rigour thought that I should help with the solution of E4.3 by proving that

(*) *if X and Y are independent random variables and X has continuous distribution function F, then* $\mathsf{P}(X = Y) = 0$.

Proof. By replacing X and Y by $h(X)$ and $h(Y)$, where $h(x) = x/(1 + |x|)$, we may – and *do now* – assume that $|X| < 1$, and $|Y| < 1$.

Let $\epsilon > 0$ be given. Then, for $n \in \mathbf{N}$,

$$\mathsf{P}(X = Y) \leq \sum_{k=-n}^n \mathsf{P}\big(k/n < X \leq (k+1)/n; \ k/n < Y \leq (k+1)/n\big)$$

$$= \sum_{k=-n}^n \big\{F((k+1)/n) - F(k/n)\big\}\mathsf{P}\big(k/n < Y \leq (k+1)/n\big).$$

But F is *uniformly* continuous on $[-1,1]$, so that we can find n such that for every k, we have $F((k+1)/n) - F(k/n) < \epsilon$. Then

$$P(X = Y) \leq \epsilon \sum_{k=-n}^n \mathsf{P}\big(k/n < Y \leq (k+1)/n\big) \leq \epsilon.$$

References

Aldous, D. (1989), *Probability Approximations via the Poisson Clumping Heuristic*, Springer, New York.

Athreya, K.B. and Ney, P. (1972), *Branching Processes*, Springer, New York, Berlin.

Billingsley, P. (1968), *Convergence of Probability Measures*, Wiley, New York.

Billingsley, P. (1979), *Probability and Measure*, Wiley, Chichester, New York (2nd edn. 1987).

Bollobás, B. (1987), Martingales, isoperimetric inequalities, and random graphs, *Coll. Math. Soc. J. Bolyai* 52, 113-39.

Breiman, L. (1968), *Probability*, Addison-Wesley, Reading, Mass..

Chow, Y.-S. and Teicher, H. (1978), *Probability Theory: Independence, Interchangeability, Martingales*, Springer, New York, Berlin.

Chung, K.L. (1968), *A Course in Probability*, Harcourt, Brace and Wold, New York.

Davis, M.H.A. and Norman, A.R. (1990), Portfolio selection with transaction costs, *Maths. of Operation Research* (to appear).

Davis, M.H.A. and Vintner, R.B. (1985), *Stochastic Modelling and Control*, Chapman and Hall, London.

Dellacherie, C. and Meyer, P.-A. (1980), *Probabilités et Potentiel*, Chaps. V–VIII, Hermann, Paris.

Deuschel, J.-D. and Stroock, D.W. (1989), *Large Deviations*, Academic Press, Boston.

Doob, J.L. (1953), *Stochastic Processes*, Wiley, New York.

Doob, J.L. (1981), *Classical Potential Theory and its Probabilistic Counterpart*, Springer, New York.

Dunford, N. and Schwartz, J.T. (1958), *Linear Operators: Part I, General Theory*, Interscience, New York.

Durrett, R. (1984), *Brownian Motion and Martingales in Analysis*, Wadsworth, Belmont, Ca.

Dym, H. and McKean, H.P. (1972), *Fourier Series and Integrals*, Academic Press, New York.

Ellis, R.S. (1985), *Entropy, Large Deviations, and Statistical Mechanics*, Springer, New York, Berlin.

Ethier, S.N. and Kurtz, T.G. (1986), *Markov Processes: Characterization and Convergence*, Wiley, New York.

Feller, W. (1957), *Introduction to Probability Theory and its Applications, Vol.1*, 2nd edn., Wiley, New York.

Freedman, D. (1971), *Brownian Motion and Diffusion*, Holden-Day, San Francisco.

Garsia, A. (1973), *Martingale Inequalities: Seminar Notes on Recent Progress*, Benjamin, Reading, Mass.

Grimmett, G.R. (1989), *Percolation Theory*, Springer, New York, Berlin.

Grimmett, G.R. and Stirzaker, D.R. (1982), *Probability and Random Processes*, Oxford University Press.

Hall, P. (1988), *Introduction to the Theory of Coverage Processes*, Wiley, New York.

Hall, P. and Heyde, C.C. (1980), *Martingale Limit Theory and its Application*, Academic Press, New York.

Halmos, P.J. (1959), *Measure Theory*, Van Nostrand, Princeton, NJ.

Hammersley, J.M. (1966), Harnesses, *Proc. Fifth Berkeley Symp. Math. Statist. and Prob., Vol.III*, 89-117, University of California Press.

Harris, T.E. (1963), *The Theory of Branching Processes*, Springer, New York, Berlin.

Jones, G. and Jones, T. (1949), (Translation of) *The Mabinogion*, Dent, London.

Karatzas, I. and Schreve, S.E. (1988), *Brownian Motion and Stochastic Calculus*, Springer, New York.

Karlin, S. and Taylor, H.M. (1981), *A Second Course in Stochastic Processes*, Academic Press, New York.

Kendall, D.G. (1966), Branching processes since 1873, *J. London Math. Soc.* **41**, 385-406.

Kendall, D.G. (1975), The genealogy of genealogy: Branching processes before (and after) 1873, *Bull. London Math. Soc.* **7**, 225-53.

Kingman, J.F.C. and Taylor, S.J. (1966), *Introduction to Measure and Probability*, Cambridge University Press.

Körner, T.W. (1988), *Fourier Analysis*, Cambridge University Press.

Laha, R. and Rohatgi, V. (1979), *Probability Theory*, Wiley, New York.

Lukacs, E. (1970), *Characteristic Functions*, 2nd edn., Griffin, London.

Meyer, P.-A. (1966), *Probability and Potential* (English translation), Blaisdell, Waltham, Mass.

Neveu, J. (1965), *Mathematical Foundation of the Calculus of Probability* (translated from the French), Holden-Day, San Francisco.

Neveu, J. (1975), *Discrete-parameter Martingales*, North-Holland, Amsterdam.

Parthasarathy, K.R. (1967), *Probability Measures on Metric Spaces*, Academic Press, New York.

Rogers, L.C.G. and Williams, D. (1987), *Diffusions, Markov Processes, and Martingales, 2: Itô calculus*, Wiley, Chichester, New York.

Ross, S. (1976), *A First Course in Probability*, Macmillan, New York.

Ross, S. (1983), *Stochastic Processes*, Wiley, New York.

Stroock, D.W. (1984), *An Introduction to the Theory of Large Deviations*, Springer, New York, Berlin.

Varadhan, S.R.S. (1984), *Large Deviations and Applications*, SIAM, Philadelphia.

Wagon, S. (1985), *The Banach-Tarski Paradox*, Encyclopaedia of Mathematics, Vol. 24, Cambridge University Press.

Whittle, P. (1990), *Risk-sensitive Optimal Control*, Wiley, Chichester, New York.

Williams, D. (1973), Some basic theorems on harnesses, in *Stochastic Analysis*, eds. D.G. Kendall and E.F. Harding, Wiley, New York, pp.349-66.

Index

(Recall that there is a Guide to Notation on pages xiv-xv.)

ABRACADABRA (4.9, E10.6).

adapted process (10.2): Doob decomposition (12.11).

σ-algebra (1.1).

algebra of sets (1.1).

almost everywhere = a.e. (1.5); almost surely = a.s. (2.4).

atoms: of σ-algebra (9.1, 14.13); of distribution function (16.5).

Azuma-Hoeffding inequality (E14.2).

Baire category theorem (A1.12).

Banach-Tarski paradox (1.0).

Bayes' formula (15.7–15.9).

Bellman Optimality Principle (E10.2, 15.3).

Black-Scholes option-pricing formula (15.2).

Blackwell's Markov chain (E4.8).

Bochner's Theorem (E16.5).

Borel-Cantelli Lemmas: First = BC1 (2.7); Second = BC2 (4.3); Lévy's extension of (12.15).

Bounded Convergence Theorem = BDD (6.2, 13.6).

branching process (Chapter 0, E12.1).

Burkholder-Davis-Gundy inequality (14.18).